Community Participation and Sustainable Forest Development: Global Perspective

Community Participation and Sustainable Forest Development: Global Perspective

Arvind Kumar

RANDOM

RANDOM PUBLICATIONS
NEW DELHI (INDIA)

Community Participation and Sustainable Forest Development: Global Perspective

---

ISBN 978-93-5111-827-5

Published in 2016 in India by

RANDOM PUBLICATIONS

4376-A/4B, Gali Murari Lal, Ansari Road
New Delhi-110 002
Phone : +9111-43580356, 011-23289044, 011-43142548
e-mail: sales@randompublications.com,
info@randompublications.com, randomexports@gmail.com

Reprinted 2022

Type Setting by : Friends Media, Delhi-110089
Digitally Printed at: Replika Press Pvt. Ltd.

Preface

Community forestry is defined as forestry by the people of the people, for the people. The people are qualified to mean the involvement of the people rather than the ownership. It is something like rehabilitation of forests for rehabilitation of people through active involvement and participation of the community the government acting as a catalyst and also a partner.

Developing effective forestry strategies and policies to promote sustainable development involves an array of difficult choices. For example, while we know that forest clearing for crops and pasture, overcutting for fuelwood, uncontrolled commercial logging for timber and expanding infrastructure all contribute to deforestation and degradation, the fundamental problem facing policy-makers is how to address the underlying causes. These include poverty, hunger, access to land, a lack of jobs and income-generating opportunities, and growing economic demands for forest goods and services.

The large and growing amount of forest owned and administered by communities, the growing recognition that communities can often be as good if not better managers of forests than governments or companies, and the potential for forest resources to contribute to sustainable community development, have all moved indigenous peoples' forestry to the center of the global forest debate. While indigenous peoples' forestry may have moved to the center, indigenous peoples themselves are not yet central players and there is a tremendous lack of awareness on barriers and market opportunities at the national and international levels, in industry, research and development organization circles. There is also a great need to facilitate new networks of communities, companies, policy-makers, investors, environmental groups and researchers. Indigenous peoples' forests and forestry present a huge challenge to the global forest community.

The book examines the phenomenon of community participation, sustainable forest management and development with reference to the forest policies and forest acts of different countries.

– Author

Contents

1

Introduction

The National Commission on Agriculture in India introduced the concept of Social Forestry to encourage those who were dependent on fuelwood, fodder and other forest products, to meet their own needs through various activities, in order to reduce the burden on the Forests. This concept was further refined by FAO in 1978, by defining community forestry as the programme which intimately involved local people in afforestation, irrespective of the pattern of land ownership. While the traditional forestry covered the protection and production roles, social forestry was intended to play the social role.

Thus community forestry programmes included a wide range of activities such as growing trees on farm bunds and roadsides, developing woodlots on common properties and collection, processing and management of forest products by involving local communities. Realising the problems of poverty and unemployment closely linked with denudation of natural resources, scarcity of fuel, fodder and timber, the Government of India in their Sixth Five Year Plan (1980-85) aimed at generating rural employment through social forestry and thereby reducing poverty from 48 per cent to 30 per cent, which was further intended to be brought down to 10 per cent during the Seventh Five Year Plan (1985-90). It was also intended to produce adequate fodder and fuel to meet the needs of the poor, particularly in forest deficit areas.

The major focus was on the extension of forestry programme to non-forest lands with the active involvement of local communities. This was essential because only about 40 per cent of the 5 million villages had access to forests, while the rest had to look for alternate sources of woodfuel. It was also an opportunity to develop 175 million ha of degraded lands. Subsequently, in the 1980s, the scope of social forestry was enlarged to cover the protection of reserved forests and establish tree species on vacant sites in urban areas. However, the programme did not include large-scale industrial forestry.

In 1985, the National Wastelands Development Board was established by the Government of India to promote the production of fodder, fuel and minor timber on wastelands, by involving local communities and voluntary agencies. Important social forestry models formulated to achieve these goals were -

rehabilitation on degraded lands, strip plantations, village woodlots, farm forestry, agroforestry, homestead plantations and decentralised nurseries. Newly established State Social Forestry Departments were confident of implementing this programme, in view of their strong technical expertise and adequate resources. However, in the absence of social mobilisation, involvement of the local people was confined only to wage earning on community woodlots, while the implementation of these schemes was carried out by the Forest Officials.

Community Forestry Schemes in 1980's

By 1990, the drawbacks of various social forestry schemes launched in different parts of the country became apparent. The Social Forestry Departments had great difficulty in mobilising local participation. There was a wide gap between the assumed problems and actual realities in the field. As the work plans were not based on specific ground data and ecological conditions, the proposed models of social forestry were often found to be inappropriate. However, modifications in the project designs and implementation plans could not be carried out because of rigid rules and regulations. These schemes could not register the anticipated success because of the following reasons:

- Community Woodlots: Development of fuelwood plantations on community wastelands was a major programme to generate employment for the local poor and the landless, while augmenting the fuel and forage shortfall faced by the community. However, this scheme could cover only 9 per cent of the villages either due to non-availability of wastelands or unwillingness of the Village Panchayats and public institutions to spare their wastelands for community plantations. Many of these projects had failed to progress in terms of plant survival, growth and income generation. In the absence of effective local organisations, the programme heavily depended on the Village Panchayats, who had bare minimum staff to manage these plantations. As the area available for plantations was very small, the supply of fuel and fodder was also inadequate to meet the local demands. It was also realised that the problems of fuelwood and fodder shortages could not be solved simply by promoting the cultivation of these species, as these commodities were of least value as compared to timber wood. Therefore, despite the shortage of fuelwood, the farmers preferred to grow timber and fruit species.

In some successful community plantations, inspite of the offer to collect grass free of cost, the response from farmers was very poor. The villagers felt that feeding such fodder to poor quality animals would not improve the productivity except in a few locations where they had maintained livestock of improved breeds. As the community plantation activities could not motivate the local people, most of these plantations were destroyed within a short period after completion of the project.

- Strip Plantations: Saplings of tall growing tree species were established in multiple lines along the railway tracks, canal bunds and roadsides and the landless families were involved in protecting them. This was fairly successful but the incentives offered by the project implementing agency were not adequate for the beneficiaries to devote sufficient time for protection. Thus some of these trees were stolen. In the absence of clear directives from the government regarding the sharing of benefits, the participants were not allowed to cut the trees.
- Farm Forestry: It was the most successful scheme wherein farmers established plantations of eucalyptus, casuarina, poplar, teak, etc., on their agricultural lands using necessary inputs like irrigation and fertilizers with the hope of finding a suitable market and attractive price for the produce. Many of these farmers located near the wood based industries could sell wood at remunerative prices. However in the absence of any marketing infrastructure it was difficult for them to find attractive buyers in remote areas. Obsolete Tree Acts imposed by the government created further obstacles in felling, transporting and sawing of timber by the growers. This forced them to either sell it to the middlemen or delay the harvest.

The community forestry programme also included a component of planting fruit species on marginal lands owned by poor families belonging to scheduled castes and scheduled tribes. However, the project could not get adequate response at the initial stage as most of the illiterate target families feared that the government would acquire their lands, after the establishment of trees. Subsequently, even the limited number of participants in the programme could not maintain the plants due to inadequate resources to provide irrigation and protection.

It was also observed that while promoting farm forestry, farmers were not informed about the profitability of various tree species which could be grown under local agro-climatic conditions. As a result, the farmers preferred only a few species such as eucalyptus, which they thought to be most profitable, while many other fruit and non-timber forest product species which could have provided higher income were neglected.

- Decentralised nurseries: The primary objective was to build up the local capacity for assured supply of planting materials of popular tree species. As the focus was on fodder and fuelwood plantations, the participants were advised to raise those saplings which were of no interest to them. In the absence of well planned marketing arrangements, the nurserymen were totally dependent on the Social

Forestry Department for selling their seedlings. As a result, farmers had to discontinue their activities immediately after completion of the project.

- Energy Conservation: Several energy conservation measures like promotion of improved wood stoves and biogas plants did not achieve the expected results due to poor publicity and awareness. The poor response was also because of unsuitable designs of smokeless wood stoves which failed to save wood.

Lessons Learnt

No doubt the community forestry programmes in 1980s could not prove to be people-oriented, but they did generate a lot of awareness among the local people. Many success stories were recorded at micro-level which were good for replication and motivation of the rural people on a larger scale. Several lessons were learnt by the Project Implementing Agencies.

Although biomass was the major source of fuel for cooking, the supply was not critical. Contrary to the earlier belief, it has now been understood that only 30 - 40 per cent of biofuel was in the form of wood and only 10 - 15 per cent of it was from natural forests. The rest of the biofuel was from crop residues, animal dung, shrubs and bushes grown on non-forest lands. However, as the demand for woodfuel will increase at 1.5 - 2.0 per cent per annum, the estimated consumption of woodfuel in 2010 will be 226 million tons against the supply of 256 million tons. Apart from the wood for fuel, the annual demand for wood from industries, housing and agricultural sectors is 125 million tons, leading to further pressure on forest resources.

Compared to other sources of energy, woodfuel has several advantages. Firstly, wood can be easily regenerated through the establishment of energy plantations on marginal and wastelands. With its tropical climate, fairly good quality soil and well distributed rainfall, India presents an ideal condition to produce 10-15 tons of biomass per ha/year. Woodfuel production and distribution can also provide employment for rural people at various levels. The woodfuel business can provide employment for about 10 per cent of the rural households and contribute to about 40 per cent of their total earning. Inspite of huge involvement of labour, energy from wood is cheaper than fossil fuels.

Woodfuel is also environment friendly, when the rate of biomass harvest is equivalent to the rate of regrowth. In such a situation, CO_2 released from wood while burning will be absorbed by the greenery without any burden on the environment. The use of woodfuel in the Asia Pacific Region in 1994 has resulted in saving 278 million tons of CO_2, which otherwise would have been released into the global atmosphere. With continued use of wood as a hypothetical replacement for LPG, the annual CO_2 saved will be 349 million tons in 2010. The estimated cost avoided by recapturing the CO_2 is estimated at US $ 14 billion in 1994 and US $ 17.5 billion in 2010. Thus, biomass production deserves high priority in community forestry. Realising this need, emphasis was laid on Joint Forest Management which involved people's organisations to launch tree plantations both on forest and non-forest lands. This programme emphasized on the formation of Village Forest Committees and capacity building of the committee members to take suitable decisions on afforestation and its protection without any interference from the donor agencies and Forest Department. Inspite of its good intentions, the success of the Joint Forest Management Programmes was also dependent on other factors such as productivity of the plantations based on soil fertility and composition of the tree species, demand for the produce, political support and transparency of the project implementing agency. In some states where the government had issued clear directives on the sharing of benefits between the government and local people, the latter had shown greater enthusiasm, resulting in a higher degree of success.

The success of the community forestry programme was also influenced by political commitment, correct assessment of local needs, appropriate technology, system of providing incentives, suitable local organisations and support services for finance, management, research and extension. Inspite of major setbacks, the Government of India has emphasized that those community forestry programmes which aim at conserving the natural resources should address the problem of poverty. It was realised that the community forestry could not solve all the problems of the rural communities, although it had good potential to support the rural economy. There was a need to involve the local communities for planning the programme till they developed suitable interventions to fulfill the goals.

The key task for successful implementation of the programme in the future was to understand the socio-economic aspects of the local population and design a suitable technical programme to ensure sustainable development. The programme should have a strong component of extension for motivating the participants. The extension programme should highlight the benefits of various schemes through effective media, such as radio, TV, documentary films and newspapers. However, as most of the farmers were illiterate, involvement of local leaders, members of the village *Panchayats*, progressive farmers, school teachers and voluntary agencies was most effective in ensuring better people's participation and adoption of new technologies.

PROGRAMME PRIORITIES FOR THE FUTURE

The community forestry programmes during the next two decades should focus on the following aspects.

- Enhancing the productivity of natural resources, while augmenting the basic needs of the community;

- Empowering the local communities to initiate the process of planning and programme implementation at the microlevel to promote afforestation for income generation and ecological conservation.

The programme during the next two decades may be divided into two phases. During the first phase of 10 years, emphasis should be given to enhance the productivity, while conserving biodiversity and the environment. The second phase should consolidate the programme by promoting micro and macro-level enterprises, involving the local people and the business interests to sustain the development and optimise the opportunities for employment and income generation.

AGROFORESTRY FOR IMPROVING PRODUCTIVITY OF PRIVATE LANDS

As over 70 per cent of the agricultural lands in India are dependent on rainfall, the crop yields are comparatively low. In such a situation, promotion of tree based farming system has good potential to improve the Land Equivalent Ratio (LER) as well as crop yields. The system of combining horticultural crops with fodder, fuel, NWFP and agricultural crops can ensure the supply of essential commodities while generating cash surplus. Among these tree species, the income from fruit and NWFP plantation are generally higher than timber and industrial wood plantation, because of substantial annual income without cutting the trees.

Development of tree based farming systems on the watershed basis has the potential to improve water supply, food production, employment generation apart from supplying fuel and forage. Water conservation can be facilitated through contour bunding, gully plugging, formation of farm ponds and percolation tanks. This programme can help in recharging the ground water table to provide water for human beings, livestock and crops. The selection of a suitable farming system will be based on the soil fertility and moisture availability. As the Government of India has already launched a massive programme of watershed development, linkage of agroforestry as an integral part of this programme will not only provide an opportunity to enhance biomass production, but also improve the profitability.

While fruit orchards are ideal for deep soils having some water resources to protect the plants in the initial stages, farm forestry and agri-silviculture are suitable for marginal lands. With introduction of fruit species, farmers can be motivated to intensify their farming operations and improve the soil productivity. Through soil and water conservation, green manuring vermi-composting, agroforestry and integrated crop protection, the yields of intercrops can be substantially high, atleast for the next 10-15 years, till the intercrops are affected by the shade, although the LER will remain significantly high to maintain the orchards. Such tree based farming systems have been highly effective in preventing migration of rural families and supporting livestock husbandry to enhance their income by 35-40 per cent.

Wood Production and Conservation

Commercial Plantations: Wood production should continue to be the priority under community forestry. However, fuelwood being a low value commodity, farmers prefer to grow superior quality wood suitable for timber and industrial use. Therefore the priority should be to promote cultivation of timber and industrial wood on degraded agricultural lands and community wastelands, which are not suitable for cultivation of fruit and NWFP species. The wood based industrial and housing sectors which are dependent on natural forests can associate with farmers to procure their raw materials. This programme can be managed with least financial support from the government, as commercial banks and industries are prepared to provide necessary finance.

Mixed Plantations: While the industries utilise the wood of a certain minimum diameter, almost 40-50 per cent biomass in the form of branches and twigs can be available for fodder, manure and fuel, almost free of cost. Apart from producing fuelwood as a byproduct in commercial plantations, fast growing fuelwood species of short gestation can be introduced on field bunds and as mixed crops with commercial wood species. This will augment the fuelwood supply to a great extent.

Wood Saving Devices: Efforts are also needed to introduce various energy conservation measures. These may include the development of fuel efficient, low cost wood stoves, promotion of improved cooking vessels and introduction of efficient models of biogas to substitute woodfuel for cooking.

Biomass being a renewable form of energy, it can be an excellent source for generating electricity in wood surplus countries. However, it is necessary to develop efficient gasifiers to take advantage of this technology in the future.

Organisational Development

Development of local organisations is the key to ensure people's participation and to sustain the benefits of investments and efforts made in community forestry.

1. Tree Growers Organisations: The initial need is to provide leadership to organise local people in afforestation. This can be carried out through formation of local Selp-Help Groups, village level farmers' associations and forest groups. This activity should receive priority during the next decade.
2. Capacity Building: Training of local farmers, supply of critical inputs, credit and technical advisory services to improve the profitability and developing market networks are the critical inputs which have to be introduced in the field to support tree growers.

3. Networking of Tree Growers and User Groups: There is an urgent need to set up networks of tree growers, forestry users and concerned scientists in these fields. Such networks should identify the opportunities for tree growers. They can also promote processing and marketing of forestry products and protect the interests of tree growers.
4. Forestry Information Service: Setting up of information centres to identify the priority of the industries and farmers and provide necessary information to enhance the economic viability of community forestry would help to sustain the interest of tree growers. Forestry Extension Units should closely associate with Information Service Centres to identify appropriate technologies for transferring to the field.
5. Forestry Research: The networks should identify the problems affecting the production and marketing of the produce and encourage the Forestry Research Institutions to take up relevant scientific studies to solve the problems. The researchers should closely associate with the network and Information Centres to disseminate new technologies to solve field problems and enhance the production.

THE DEVELOPMENT OF FORESTRY

A feature of fundamental importance, which distinguishes forests from some other resources, is that they are renewable. This means that the mature forest, as it is cut, may be replaced by a new crop of trees; thus a continuous yield of wood is provided from the same area of land. The application of this principle is the primary basis of forestry practice, which may be defined briefly as the growing and using of successive crops of timber.

As the practice of forestry develops, the forests often come to be looked upon less as a natural resource and more as a crop that is dependent in large measure on man's knowledge and skill. However, forestry is not concerned exclusively with growing timber as a crop; forests contribute in many other ways to the well-being of mankind and proper forestry practices assure the maintenance and enlargement of these benefits. Man is not a conservationist by nature: history shows that as long as the existing forest appears to be adequate to the demand, it is subject to exploitation for immediate revenue and there is little concern for its future productivity. In other words, forest conservation, involving some investment of time and money, only begins when it is apparent that current practices are endangering the resource with critical effects upon the supply of forest products and services.

Down through the centuries, as man has increased in numbers and spread over the face of the earth, he has caused and is still causing widespread destruction of the forest cover, partly of necessity to provide land for agriculture, settlement and industrial development, partly through ignorance and greed and partly in the belief that the forest resource is inexhaustible.

This latter viewpoint is to some extent justified in the early development of countries and regions that have extensive forest wealth and relatively small populations. Here the forest is often looked upon as a hindrance to settlement and agricultural development, with the result that areas cleared may exceed requirements and include lands quite unsuitable for the purpose in mind.

Man has in fact retained under intensive agriculture only a small portion of the area that was originally cleared for this purpose. It is estimated that, as a result of his activities, man has destroyed about two thirds of the world's original forest cover; the remaining forests containing merchantable timber today occupy less than one fifth of the world's land area.

In some parts of the world forest destruction has been unchecked, despite the adverse effect that this has had on the general economy and on the living standards of the people. Asia, in particular, has suffered from the effects of forest clearance and forest destruction, with widespread soil deterioration, inadequate supplies of wood to meet basic domestic needs and low living standards for a large portion of the population. Elsewhere, as the danger became apparent, methods of managing the forests to ensure their continued productiveness were evolved and thus the science of forestry came into being. However, the philosophy of conservation has developed slowly and forestry has invariably been the child of necessity: only after periods of forest destruction and land clearance has the need for husbandry and planned forest economy been realized.

TRANSITION FROM EXPLOITATION TO MANAGEMENT

The transition from exploitation to management takes time and requires the solution of many problems, technical, economic and administrative. In the beginning it may be characterized by a strong reaction against spoliation aimed at preserving the forest intact — "locking it up" — for posterity. In general, however, this is the antithesis of sound management of a renewable resource, which for its fullest development requires periodic harvesting and regeneration; an exception would be the preservation of representative samples of natural forest for scientific study and as "wilderness" areas. Modern forestry originated in Germany and France and much of its development during the past two centuries has taken place in northwestern Europe. Attention to forest conservation in these countries came early owing to an increasing demand for timber and firewood and the difficulty of securing wood cheaply from other countries.

These conditions and the serious effects of deforestation on stream flow, particularly in montane regions, made evident an urgent need for forest management. During the nineteenth century expanding populations and the iron and coal economy of the industrial revolution created new demands for wood and gave added impetus to the development of forestry.

Slowly and usually only as the need became urgent, other countries have followed the lead of western Europe and developed forestry practices to meet their own requirements. Such developments usually begin with restriction on the use of the forest, as in the setting aside of timber reserves and with protection of the forest against its more serious enemies, particularly fire. These steps are followed by efforts to secure new growth after cutting, by encouraging natural reproduction and by planting: with this the practice of silviculture begins. Finally, the accessible, productive forests are managed under cutting regulations and suitable silvicultural techniques to ensure a continuous yield of timber in perpetuity. Such forests are said to be managed on a "sustained-yield" basis.

While the natural forest usually provides the starting point for a forestry programme, it may not represent the optimum conditions that an area is capable of and its perpetuation is not always looked upon as the ideal objective of forest management. The aim, rather, may be to improve the natural forest as far as possible within the limits set by the local environment, silvicultural knowledge and economic considerations and in view of the purpose for which the forest is to be used.

Attempts to go beyond these limits and completely change the natural forest pattern have sometimes failed. A well-known instance was the century-long attempt to convert the lowland hardwood forests of Saxony to pure Norway spruce.

Despite early rapid growth, the spruce was often badly damaged by snow and frost as well as by fungal and insect attacks. By the end of the third crop of spruce, conditions had so deteriorated that hardwood species had to be reintroduced and it was generally conceded that the plan had been unsuccessful. On the other hand, there are many examples of the successful introduction of types and species of trees different from those in the natural forest. The success of these operations depends largely on careful study of the ecological factors involved.

The aim of forestry may vary from one country to another and even between regions within the same country, depending upon local conditions. Usually the primary purpose is to grow sufficient wood to meet a country's requirements — first for domestic consumption, then for export to foreign markets. Products other than wood, usually referred to as minor forest products, may sometimes be an objective of management. These vary in importance and kind from one country to another, *e.g.*, turpentine, resin, oils, fruits, maple syrup, bark for tanning, etc. The southeastern United States and the Landes in Gascony, France, are two regions where the production of resin and turpentine from species of pine is a major industry carried on in association with timber production.

While the production of wood and other commodities is often primary, in its broader aspects forestry is concerned with the interaction between the forest and its environment, as, for example, in the effects of the forest on the stabilization of soil, on the maintenance and restoration of soil fertility, on the regulation of stream flow and on the provision of a suitable environment for valuable fish and wild life. The protective influences of the forest are many and far-reaching and under certain conditions may be more important than the supplying of wood.

This is particularly true in some parts of the tropics and subtropics, in crucial drainage areas and in regions of light, sandy soil unsuitable for cultivation, where the removal of the forest cover often results in near-desert conditions. Forestry may also aim at the full development of the recreational value of an area; but even where recreation or protection are the main objectives, controlled cutting of wood can usually be carried on advantageously, if the cutting operations do not interfere with the primary purpose of management.

The term "multiple use" is widely used in referring to the management of a forest area for two or more purposes, one of which may be dominant, but not exclusive.

Bases of Forestry

Forestry may be defined as the scientific management of forests for the continuous production of goods and services. While its practice involves many problems of an economic and administrative nature, the solution of which is essential for success, its proper development must in the first place be based on a sound knowledge of the forests and of the laws that govern their development and on an understanding of the characteristics of the individual species that make up the forests and of the interrelationship of the forests with their environment.

The forests of the world include a great many species of trees, not necessarily having any close botanical relationship, with varying adaptations to climate, soil and topography and more or less distinct reproductive characteristics, forms and rates of growth, sizes and ages at maturity, light requirements and abilities to resist damage by insects and other destructive agencies. It is with an intimate knowledge of these features of the trees and tree communities that sound forestry practice must begin.

SUPPLY OF TIMBER : IMPORTS AND EXPORTS

The following table shows the net imports and exports of European countries (average data, calculated from the returns of recent years). The only timber-exporting countries of Europe are Russia, Sweden, Norway, Austria-Hungary and Rumania; all the others either have only enough for their own consumption, or import timber. Great Britain and Ireland import now upwards of 10,000,000 tons a year, Germany about 4,600,000 tons and Belgium about 1,300,000 tons.

Holland, France, Portugal, Spain and Italy are all importing countries, as also are Asia Minor, Egypt and Algeria. The west coast of Africa exports hardwoods and imports coniferous timber. The Cape and Natal import considerable quantities of pine and fir wood. Australasia Net Imports and Exports of European Countries. These net imports are received from non-European countries. They consist chiefly of valuable hardwoods, like teak, mahogany, eucalypts and others. Exports hardwoods and some Kauri pine from New Zealand, but imports larger quantities of light pine and fir timber.

British India and Siam export teak and small quantities of fancy woods. The West Indies and South America export hardwoods and import pine and fir wood. The United States of America will not much longer be a genuine exporting country, since they import already almost as much timber from Canada as they export. Canada exports considerable quantities of timber. The Dominion has still a forest area of 1,250,000 sq. m., equal to 38% of the total area and giving 165 acres of forest for every inhabitant. Although only about one-third of the forest area can be called regular timber land, Canada possesses an enormous forest wealth, with which she might supply permanently nearly all other countries deficient in material, if the governing bodies in the several provinces would only determine to stop the present fearful waste caused by axe and fire and to introduce a regular system of management. As matters stand, the supplies of the most valuable timber of Canada, the white or Weymouth pine, are nearly exhausted, the great stores of spruce in the eastern provinces are being rapidly destroyed and the forests of Douglas fir in the western provinces have been attacked for export to the United States and to other countries.

Taking the remaining stocks of the whole earth together, it may be said that a sufficient quantity of hardwoods is available, but the only countries which are able to supply coniferous timber for export on a considerable scale are Russia, Sweden, Norway, Austria and Canada. As these countries have practically to supply the rest of the world and as the management of their forests is far from satisfactory, the question of supplying light pine and fir timber, which forms the very staff of life of the wood industries, must become a very serious matter before many years have passed.

Unmistakable signs of the coming crisis are everywhere visible to all who wish to see and it is difficult to over-state the gravity of the problem, when it is remembered, for instance, that 87% of all the timber imported into Great Britain consists of light pine and fir and that most of the other importing countries are similarly situated.

In some of these countries little or no room exists for the extension of woodland, but this statement does not apply to Great Britain and Ireland, which contain upwards of 12,000,000 acres of waste land and 12, 500,000 acres of mountain and heath land used for light grazing. Onefourth of that area, if put under forest, would produce all the timber now imported which can be grown in Britain, that is to say, about 95% of the total.

Forest Managemet

In early times there was practically no forest management. As long as the forests occupied considerable areas, their produce was looked upon as the free gift of nature, like air and water; men took it, used it and even destroyed it without let or hindrance. With the gradual increase of population and the consequent reduction of the forest area, proprietary ideas developed; people claimed the ownership of certain forests and proceeded to protect them against outsiders. Subsequently the law of the country was called in to help in protection, leading to the promulgation of special forest laws. By degrees it was found that mere protection was not sufficient and that steps must be taken to enforce a more judicious treatment, as well as to limit the removal of timber to what the forests were capable of producing permanently.

The teaching of natural science and of political economy was brought to bear upon the subject, so that now forestry has become a special science. This is recognized in many countries, amongst which Germany stands first, closely followed by France, Austria, Denmark and Belgium. Of non-European countries the palm belongs to British India and then follow Ceylon, the Malay States, the Cape of Good Hope and Japan.

The United States of America have also turned their attention to the subject. Most of the British colonies are, in this respect, as yet in a backward state and the matter has still to be fought out in Great Britain and Ireland, though many writers have urged the importance of the question upon the public and the government. There can be no doubt that all civilized countries must, sooner or later, adopt a rational and systematic treatment of their forests.

For details as to the separate countries, see the articles under the country headings; in this article only some of the more important countries are dealt with, in so far as the history of their forestry is important. A few notes on Germany and France will be given, because in these countries forest management has been brought to highest perfection; Italy is mentioned, because she has allowed her forests to be destroyed; and a short description of forestry in the United Kingdom and in India follows.

The winters being long and severe, an abundant supply of fuel is almost as essential as a sufficient supply of food. This necessity has led, along with a passion for the chase, to the preservation of forests and to the establishment of an admirable system of forest cultivation, almost as carefully conducted as field tillage. The Black Forest stretches the whole length of the grand-duchy of Baden and part of the kingdom of Wurttemberg, from the Neckar to Basel and the Lake of Constance. The vegetation resembles that of the Vosges; forests of spruce, silver fir, Scotch pine and, mingled with birches, beech and oak, are the chief woods met with. Until comparatively recent times large quantities of timber derived from these forests were floated down the Rhine to Holland and also shipped to England.

Now the greater part of it is used locally for construction, or it is converted into paper pulp. In the grand-duchy of Hesse the Odenwald range of mountains, stretching between the Main and the Neckar, contains the chief supply of timber. In the province of Nassau there are the large wooded tracts of the Taunus mountain range and the Westerwald.

In Rhenish Prussia valuable forests lie partly in the Eifel, on the borders of Belgium and on the mountains overhanging the Upper Moselle, but they do not furnish such stately trees as the Black Forest and the Odenwald. The Spessart, near Aschaffenburg in Bavaria, is one of the most extensive forests of middle Germany, containing large masses of fine oak and beech, with plantations of coniferous trees, such as spruce, Scotch pine and silver fir.

Bavaria possesses other fine forest tracts, such as the Baierischewald on the Bohemian frontier, the Kranzberg near Munich and the Frankenwald in the north of the kingdom. North Germany has extensive forests on the Harz and Thuringian Mountains, while in East Prussia large tracts of flat ground are covered with Scotch pine, spruce, oak and beech.

Every German state has its forest organization: In Prussia the department is presided over by the Oberland Forstmeister at Berlin, while each province, or part of a province, has an Oberforstmeister, under whom a number of OberfOrsters administrate the state and communal forests. These, again, are assisted by a lower class of officials called Forsters. The Oberf6rsters throughout Germany are educated at special schools of forestry, of which in 1909 the following nine existed: In Prussia: at Eberswalde and Munden.

The National Forest Policy

The forest policy of the United States may be said to have had its origin in 1799 in the enactment of a law which authorized the purchase of timber suitable for the use of the navy, or of land upon which such timber was growing. It is true that laws were in force under the early governments of Massachusetts, New Jersey and other colonies, providing for the care and protection of forest interests in various ways, but these laws were distinctly survivals of tendencies acquired in Europe and for the most part of little use.

It was not until the apparent approach of a dangerous shortage in certain timber supplies that the first real step in forest policy was taken by the United States. Successive laws passed from 1817 to 1831 strove to give larger effect to the original enactment, but without permanent influence towards the preservation of the live oak, which was the object in view. A long period of inaction followed these early measures. *The unshaded areas are treeless, except along the Streams* 1831 the solicitor of the treasury assumed a partial responsibility for the care and protection of the public timber lands and in 18 55 this duty was transferred to the commissioner of the general land office in the Department of the Interior.

The effect of these changes upon forest protection was unimportant. When, however, at the close of the Civil War railway building in the United States took on an unparalleled activity, the destruction of forests by fire and the axe increased in a corresponding ratio and public sentiment began to take alarm. Action by several of the states slightly preceded that of the Federal government, but in 1876 Congress, acting under the inspiration of a memorial from the American Association for the Advancement of Science, authorized the appointment of an officer under the commissioner of agriculture, to collect and distribute information upon forest matters.

As the railways advanced into the treeless interior, public interest in tree-planting became keen. In 1873 Congress passed and later amended and repealed the timber culture acts, which granted homesteads on the treeless public lands to settlers who planted one-fourth of their entries with trees. Though these measures were not successful in themselves they directed attention towards forestry. The act which repealed them in 1891 contained a clause which lies at the foundation of the present forest policy of the United States.

By it the president was authorized to set aside " any part of the public lands wholly or in part covered with timber or undergrowth, whether of commercial value or not, as public reservations and the President shall, by public proclamation, declare the establishment of such reservations and the limits thereof." Some eighteen million :acres had been proclaimed as reservations at the time when, in 1896, the National Academy of Sciences was asked by the secretary of the interior to make an investigation and report upon " the inauguration of a rational forest policy for the forest lands of the United States." Upon the recommendation of a commission named by the Academy, President Cleveland established more than twenty-one million acres of new reserves on the 22nd of February 1897. His action was widely misunderstood and attacked, but it awakened a public interest in forest questions without which the rapid progress of forestry in the United States since that time could never have been made.

Within a few months after the proclamation of the Cleveland reserves the present national forest policy took definite shape. Under this policy the national government holds and manages, in the common interest of all users of the forests or its products, such portions of the public lands as have been set aside by presidential proclamation in accordance with the act of 1891.

These lands are held against private acquisition under the Homestead Act (except as to agricultural lands as hereafter mentioned), the Timber and Stone Act and other laws under which the United States disposes of its unappropriated public domain, but not against private acquisition under the Mineral Land Laws. They are selected from lands believed to be more valuable for forest purposes than for agriculture and are managed with the purpose of securing from them the best and largest possible returns, present and future, whether in the form of water for irrigation or power, of timber, of forage for stock, or of any other beneficial product. The aggregate area of the reserves, or national forests, has been steadily increased until they now include nearly all the timber lands left of the public domain.

The general lines of this policy were in part laid down by the commission already mentioned, in its report submitted to the secretary of the interior, May 1, 1897 and by the act of June 4, 1897, which was largely shaped by the work of the commission. Until this act was passed the national forests had been in theory closed against any form of use; nor had the possibility of securing forest preservation by wise use received much thought from those who had favoured their creation. Such a state of affairs could not continue.

Before long public opinion would have forced the opening to use of the resources thus :arbitrarily locked up and in the absence of any administrative system providing for conservative use, the national forests would inevitably have been abolished and the whole policy of government forest holdings would have ceased.

The act of June 4, 1897 was therefore of the first importance. This act conferred upon the secretary of the interior general powers for the proper management of the national forests through the general land office of his department. It provided for the designation and sale of dead, mature and large timber; authorized the secretary to permit free use of timber in small quantities by settlers, miners and residents; empowered him to "make such rules and regulations and establish such service as will insure the objects of such reservations, namely, to regulate their occupancy and use and to preserve the forests thereon from destruction "; and made violation of the act or of such rules and regulations a misdemeanour. The statute limited the power to establish forest reservations to the purpose of improving and protecting the forest, securing favourable conditions of water flows and furnishing a continuous supply of timber for the use and necessities of citizens of the United States.

Lands found, upon due examination, to be more valuable for other purposes than for forest uses might be eliminated from any reservation and all mineral lands within the reservations were left open to private appropriation under the mineral laws. The rights of settlers and claimants were safeguarded and civil and criminal jurisdiction, except so far as the punishment of offences against the United States in the reservations was concerned, was reserved to the States.

While the administration of the national forests was entrusted to the general land office, the same act assigned the surveying and mapping of them to the United States Geological Survey, which has published descriptions and maps of some of the more important.

No attempt was made in the general land office to develop a technical forest service. There were, indeed, at the time of passage of the act, less than ten trained foresters in the United States, no means of training more and very little conception of what forestry actually meant. The purpose of the administration was therefore mainly protection against trespass and fire, particularly the latter.

Regulations were made giving effect to the provisions of the act of June 4, set forth above, but in the absence of technical knowledge as to what might safely be done, the tendency was rather to restrict than to extend the use of the forest. Meanwhile, however, there was rapidly developing in another branch of the government service an organization qualified for actual forest management.

One year after the passage of the act of June 4, 1897, the division of forestry in the Department of Agriculture ceased to be merely a bureau of information and became an active agency for introducing the actual practice of forestry among private owners and for conducting the investigations upon which a sound American forest practice could be based. The work awakened great interest among forest owners and exerted a powerful educational influence upon the country at large.

The division extended its work and became the Bureau of Forestry. It drew into its employment for a time nearly all the men who were preparing themselves in increasing numbers (at first abroad, then in the newly-founded schools in the United States) for the profession of forestry and was soon recognized as qualified to speak authoritatively on technical questions connected with the administration of the national forests.

This led to a request from the secretary of the interior for the advice of the bureau on such questions. Working plans were accordingly undertaken for a number of the forests. The general land office, however, was not ready to attempt active forest management. Though some timber was sold and the grazing of stock regulated to some extent, the main object of the land office administration continued to be protection against fire. Many of the regulations which it made could not be enforced.

The disadvantages of dispersal of the Federal government forest work among three separate agencies grew more and more apparent, until, on the 1st of February 1905, control of the 63,000,000 acres of forest reserves which up to that time had been set aside was transferred from the general land office to the Bureau of Forestry. In recognition of its new duties the designation of the bureau became the Forest Service.

Other provisions of the act which affected the transfer were that forest supervisors and rangers should be selected, so far as possible, from qualified citizens of the state or territory in which each forest was situated and that all money received from the sale of any products or the use of any land or resources of the national forests should be covered into the treasury and constitute a special fund for their protection, administration, improvement and extension.

Five days later a statute gave forest officers the power to arrest trespassers; and on the 3rd of March the lieu land selection law was repealed. This law had opened the way for grave abuses through the exchange of worthless land by private owners within the forests for an equal area of valuable timber lands outside.

The law has been modified since by the change of the old name " Forest Reserves " to " National Forests." The act of June iz, 1906, opened to homestead entry lands within national forests found by examination to be chiefly valuable for agriculture. The administration and improvement of the national forests are now provided for directly by congressional appropriation. The power to create national forests conferred on the president by the act of March 1891 has been repealed for the states of Washington, Oregon, Idaho, Montana, Wyoming and Colorado, but for no others.

The Forest Service began in earnest the development of all the resources of the national forests. Mature timber was sold wherever there was a demand for it and the permanent welfare of the forests and protection of the streams permitted, but always so as to prevent waste, guard against fire, protect young. growth and ensure reproduction. Regulations were adopted which allowed small sales to be made without formality or delay,. secured for the government the full value of timber sold and eliminated unnecessary routine.

Care was taken to safeguard the interests of the government and provide for the maintenance of good technical standards. The conduct of local business was entrusted to local officers. Large transactions with general policies were controlled from Washington, but with careful 2 = *District Boundaries and Numbers* provision for first-hand knowledge and close touch with the work in the field.

Business efficiency and the convenience of the public were carefully studied. In short, an organization was created capable of handling safely, speedily and satisfactorily the complex business of making useful a forest property of vast extent, scattered through sixteen different states of an aggregate area of over 1,50o,000 sq. m. and with a population of 9,000,000.

The growth since the 1st of July 1897 of the area of the national forests, of the expenditures of the government for forestry and of the receipts from the national forests, is shown by the statement which follows. Though the act of June 4, 1897, became effective immediately upon its passage, the fiscal year 1899 was the first of actual administration, because the first for which Congress made the appropriation necessary to carry out the law.

Forest is ripe for the axe, the demand is strong and control by trained men makes it safe to cut more freely. The increase is marked both in small and in large sales, but a score of sales for less than $5000 are made against one for more. The total cut is still far below the annual increment of the forests. As the demand grows restrictions must increase in order to husband the present supply until the next crop matures.

- The stumpage price would seem on the face of the figures to have risen from about one dollar to more than three dollars per thousand board-feet. The receipts, however, for any one year are not exclusively for the timber cut in that year, since payments are made in advance. In the year 1907 the average price obtained was something less than $2.50 per thousand. It is therefore true that stumpage prices have risen greatly, although conditions new to the American lumbermen are im *Area of National Forests, Annual Expenditures of the Federal Government for Forestry and National Forest Administration and Receipts from National Forests, 1898-1909* Until 1906, the sole source of receipts was the sale of timber. In the fiscal year 1907, however, timber sales furnished less than half the receipts. The following statement concerning the timber sales of the fiscal years 1904-1907 will serve to bring out the change that followed the transfer of control to the forest service in the midst of the fiscal year 1905:
  - A large excess each year in the amount of timber sold over that cut and paid for;
  - Nine times as much timber sold at the end of the four-year period as at the beginning and three times as much cut;
  - A much higher price obtained per thousand board-feet at the end of the period than at the beginning. Each of these matters calls for comment. The sales are of stumpage only; the government does no logging on its own account.
- More timber is sold each year than is cut and paid for, because many of the sales extend over several years. With increasing sales the amount sold each year for future removal has exceeded the amount to be removed during that year under sales of earlier years. Large sales covering a term of years are made because the national forests contain much overmature timber, which needs removal, but which is frequently too inaccessible to be saleable in small amounts. To prevent speculation the time allowed for cutting is never more than five years and cutting must begin at once and be continued steadily.
- The volume of sales has increased rapidly because much 1 The United States fiscal year ends June 30 and receives its designation from the calendar year in which it terminates. Thus, the fiscal year 1898 is the year July 1, 1897 - June 30, 1898.

Administration transferred to Bureau of Forestry, February I, 1905. Full utilization of all merchantable material, care of young growth in felling and logging and the piling of brush, to be subsequently burned by the forest officers if burning is necessary, are among these conditions.

Timber to be cut must first be marked by the forest officers. Sales of more than $100 in value are made only after public advertisement.

Only the simplest forms of silviculture have as yet been introduced. The vast area of the national forests, the comparatively sparse population of the West, the rough and broken character of the forests themselves and the newness of the problems which their management presents, make the general application of intensive methods for the present impracticable.

Natural reproduction is secured. The selection system is most used, often under the rough and ready method of an approximate diameter limit, with the reservation of seed trees where needed. The tendency, however, is strongly towards a more flexible and effective application of the selection principle, as a better trained field force is developed and as market conditions improve.

One conspicuous achievement was the reduction of loss by fires on the national forests. During the unusually dry season of 1905 there were only eight fires of any importance and the area burned over amounted only to about 16 of 1% of the total area. In 1906 about 12 of 1% was burned. This was accomplished by efficient patrol, co-operation of the public and by preventive measures, such as piling and burning the brush on cut-over areas.

Since the beginning of 1906 the largest source of income from the national forests was their use for grazing. Stock-raising is one of the most important industries of the West. Formerly cattle and sheep grazed freely on all parts of the public domain. In the early days of the national forests the wisdom of permitting any grazing at all upon them was sharply questioned.

Unrestricted grazing had led to friction between individuals, the deterioration of much of the range through overstocking and serious injury to the forests and stream flow. The forests of the West, however, are largely of open growth and contain many grassy parks, the results of old fires and many high mountain meadows.

Under proper regulations the grass and other forage plants which they produce in great quantity can be used without detriment to the forests themselves and with great benefit to the stock industry, which often can find summer pasturage nowhere else.

Except in southern California grazing is now permitted on all national forests unless the watersheds furnish water for domestic use; but the time of entering and leaving, the number of head to be grazed by each applicant and the part of the range to be occupied are carefully prescribed. Planted areas and cut-over areas are closed to stock until the young growth is safe from harm and goats are allowed only in the brushland of the foothills.

The results of regulation, in addition to the protection of forest growth and streams, are the prevention of disputes, improved range, better stock, stable conditions in the stock industry and the best use of the range in the interest of progress and development. The first right to graze stock on the forests is given to residents, small owners and those who have used the range before. Thus the crowding out of the weaker by the stronger and of the settler by the roving outsider has been stopped. In 1906 the forest service began to impose a moderate charge for the use of the national forest range.

The following statement shows the amount of stock grazed on the national forests 1904-09 and the receipts for the grazing charge: - A work of enormous magnitude which has now begun is planting on the national forests. At present, with low stumpage prices and incomplete utilization of forest products, clear cutting with subsequent planting is not practicable.

There are, however, many million acres of denuded land within the national forests which require planting. Such planting is still confined chiefly to watersheds which supply cities and towns with water. The first planting was done in 1892, in California. Since then similar work has been done on city watersheds in Colorado, Utah, Idaho and New Mexico. Other plantations are in the Black Hills national forest, where large areas of cut-over and burned-over land are entirely without seed trees and in the sandhill region of Nebraska. Up to 1908 about 2,000,000 seedlings had been planted, on over 2000 acres - a small beginning, but the work was entirely new and presented many hard problems.

The nursery operations of the forest service are concentrated at seven stations, located in southern California, Nebraska, Colorado, New Mexico Utah and Idaho, where stock is raised for local planting and for shipment elsewhere. These nurseries are small. Their annual productive capacity is between 8,000,000 and seedlings. Each nursery is practically an experimental forest-planting station, at which a large variety of species are grown and various methods are tried.

The organization of the administrative work of the national forests is by single forests. On the 1st of January 1908 the total number of forests was 165 with a total area of 162,023,190 acres. In charge of each forest is a forest supervisor. Under the supervisors are forest rangers and forest guards, whose duties include patrol, marking timber and scaling logs, enforcing the regulations and conducting some of the minor business arising from the use of the forests. Guards are temporary employes; rangers are employed by the year. The supervisors report directly to and receive instructions from the central office at Washington.

In this office there are four branches - operation, grazing, silviculture and products - each of which directs that part of the work which belongs to it, dealing directly with the supervisor.

For inspection purposes, however, the forests are separated into six districts, in each of which is located a chief inspector with a corps of assistants. The inspectors are without administrative authority, but assist by their counsel the supervisors and through inspection reports keep the Washington office informed of the condition of all lines of administrative work in progress. Administrative officers alternate frequently between field and office duties.

The number of forest officers in the several grades on the 1st of January 1908 were: 6 chief inspectors, 26 inspectors, 106 forest supervisors, 41 deputy forest supervisors, 820 forest rangers and 283 forest guards. The total number of employes of the forest service on the same date, including the clerical force, was 2034.

Besides the administration of the national forests, the forest service conducts general investigations, carries on an extensive educational work and co-operates with private owners who contemplate forest management upon their own tracts. This last work is undertaken because of the need of bringing forestry into practice, the lack of trained foresters outside of the employ of the government and the lack of information as to how to apply forestry and what returns may be obtained. Co-operation takes the form of advice upon the ground and, on occasion, of the making of working plans.

The educational work of the service is performed chiefly through publications, the purpose of which is to spread very widely a knowledge of the importance of forestry to the nation and of the principles upon which its practice rests. The investigations which the service conducts extend from studies. of the natural distribution and classification of American forests. and of their varied silvicultural problems to statistics of lumber production and laboratory researches which bear upon the economical utilization of forest products. As examples of these researches may be mentioned tests of the strength of timber, studies of the preservative treatment of wood for various uses, wood-pulp investigations and studies in wood chemistry.

Forestry Among the States

Among the states forestry has hardly reached the stage of practical application on the ground. New York holds 1,500,00o acres of forest land. It has a commission to care for its forest preserve and to protect the forest land throughout the state from fire. The constitution of the state, however, prohibits the cutting of timber on state land and thus confines the work entirely to protection of the forest and to the planting of waste areas. Pennsylvania is at present showing the most efficient activity in working out a forest policy. It has state forests of 820,000 acres, a good fire law more and more satisfactorily enforced and eight nurseries for growing planting material. In 1905, 160,000 white pine seedlings were set out. It has also a school for forest rangers, to be employed on the state forests and it has just established a state professional school of forestry.

Twenty-six of the states have regularly appointed forest officers, six have carried on studies of forest conditions in co-operation with the forest service and there is scarcely one which is not actively interested in forestry. Laws, generally good, to prevent damage from forest fires, have been enacted by practically all the states, but their enforcement has unfortunately been lax.

Public sentiment, however, is making rapid progress. Among the best laws are those of Maine, New Hampshire, Minnesota, New York, Pennsylvania and Wisconsin. The New York law, for example, provides for the appointment of one or more firewardens in each town of the counties in which damage by fire is especially to be feared. In other counties supervisors of towns are *ex-officio* fire-wardens. A chief fire-warden has general supervision of their work.

The wardens, half of the cost of whose services is paid by the state, receive compensation only for the time actually employed in fighting fires. They may command the service of any citizen to assist them. Setting fire to woods or waste lands belonging to the state or to another, if such fire results in loss, is punishable by a fine not exceeding $250 or imprisonment not exceeding one year, or both and damages are provided for the person injured. Since fire is beyond question the most dangerous enemy of forests in the United States, the measures taken against it are of vital importance.

The following table shows the amount of forest land held by the different states and by the territory of Hawaii: A *rea of State Forest Reservations, 1907. Forestry on Private Lands. The* practice of forestry among private owners is of old date. One of the earliest instances was that of Jared Eliot, who, in 1730, began the systematic cutting of timber land to supply charcoal for an iron furnace at Old Salisbury, Connecticut. The successful planting of waste lands with timber trees in Massachusetts dates from about ten years later. But such examples were comparatively rare until recent times.

At present the intelligent harvesting of timber with a view to successive crops, which is forestry, is much more common than is usually supposed. Among farmers it is especially frequent. It was begun among lumbermen by the late E. S. Coe, of Bangor, Maine, who made a practice of restricting the cut of spruce from his forests to trees ro, 12 or sometimes even 14 in. in diameter, with the result that much of his land yielded, during his life, a second crop as plentiful as the first.

Many owners of spruce lands have followed his example, but until very recently without improving upon it. Systematic forestry on a large scale among lumbermen was begun in the Adirondacks during the summer of 1898 on the lands of Dr W. S. Webb and Hon. W. C. Whitney, of a combined area of over ioo,000 acres, under the superintendence of the then Division of Forestry. In these forests spruce, maple, beech and birch predominate, but the spruce alone is at present of tht first commercial importance.

The treatment is a form of the selection system. Under it a second crop of equal yield would be ripe for the axe in thirtyfive years. Spruce and pine are the only trees cut. The work had been executed, at least up to the year 1902, with great satisfaction to the owners and the lumbering contractors, as well as to the decided benefit of the forest. The lumbering is regulated by the following rules and competent inspectors are employed to see that they are rightly carried out:

- No trees shall be cut which are not marked.
- All trees marked shall be cut.
- No trees shall be left lodged in the woods and none shall be overlooked by the skidders or haulers.
- All merchantable logs which are as large as 6 in. in diameter at the small end must be utilized.
- No stumps shall be cut more than 6 in. higher than the stump is wide.
- No spruce shall be used for bridges, corduroy, skids, slides, or for any purpose except building camps, dams or booms, unless it is absolutely necessary on account of lack of other timber.
- All merchantable spruce used for skidways must be cut into logs and hauled out.
- Contractors must not do any unnecessary damage to young growth in lumbering; and if any is done, they must discharge the men who did it.

These two instances of forestry have been most useful and effective among lumbermen and other owners of forest land in the north-east. Among those which have followed their example are the Berlin Mills Paper Company in northern New Hampshire,, the Cleveland Cliffs Iron Company in northern Michigan and the Delaware and Hudson Railroad Company in New York, all of which have employed professional foresters.

The most notable instance of forestry in the south is on the estate of George W. Vanderbilt at Biltmore, N.C. This was the first case of systematic forestry under regular working plans in the United States. It was begun in 1891 on about 4000 acres and has since been extended until it now covers about ioo,000 acres. A professional forester with a corps of trained rangers under him is in charge of the work. The Pennsylvania Railroad has recently employed a trained forester and several assistants and has undertaken systematic forestry on a large scale.

The effect of the work of the forest service in assisting private owners is evidenced by the fact that down to the year 1908 670 wood lots and timber tracts had been examined by agents of the forest service, of which 250 were tracts over 400 acres in extent and planting plans had been made for 436 owners covering a total area of 80,000 acres. Expert advice is also given to wood lot owners upon application by many of the state foresters.

American Practice

The conditions under which forestry is practised in Europe and in America differ so widely that rules which are received as axiomatic in the one must often be rejected in the other. Among these conditions in America are the highly developed and specialized methods and machinery of lumbering, the greater facilities for transportation and consequent greater mobility of the lumber trade, the vast number of small holdings of forest land and the enormous supply of low-grade wood in the timbered regions.

High taxes on forest properties, cut-over as well as virgin, notably in the north-western pineries and the firmly established habits of lumbermen, are factors of great importance. From these and other considerations it follows that such generally accepted essentials of European methods of forestry as a sustained annual yield, a permanent force of forest labourers, a permanent road system and the like, are in most cases utterly inapplicable in the United States at the present day in private forestry. Methods of forest management, to find acceptance, must there conform as closely as possible to existing methods of lumbering. Rules of marked simplicity, the observance of which will yet secure the safety of the forest, must open the way for more refined methods in the future. For the present a periodic or irregular yield, temporary means of transport, constantly changing crews and an almost total ignorance of the silvics of all but a few of the most important trees - all combine to enforce the simplest silvicultural treatment and the utmost concentration of purpose on the two main objects of forestry, which are the production of a net revenue and the perpetuation of the forest. Such concentration has been followed in practice by complete success.

The forests with which the American forester deals are rich in species,usually endowed with abundant powers of reproduction and, over a large part of their range, greatly dependent for their composition and general character upon the action of forest fires. Of the commercially valuable trees there may be said to be, in round numbers, a hundred out of a total forest flora of about 500 species, but many trees not yet of importance in the lumber trade will become so hereafter, as has already happened in many cases.

The attention of the forester must usually be concentrated upon the growth and reproduction of a single species and never of more than a very few. Thus the silvicultural problems which must be solved in the practice of forestry in America are fortunately less complicated than the presence of so many kinds of trees in forests of such diverse types would naturally seem to indicate.

The forest fire problem is one of the most difficult with which the American forester has to deal. It is probable that forest fires have had more to do with the character and distribution of forests in America than any other factor except rainfall. With an annual range over thousands of square miles, in many portions of the United States they occur regularly year after year on the same ground. Trees whose thick bark or abundant seeding gives them peculiar powers of resistance, frequently owe their exclusive possessions of vast areas purely to the action of fire.

On the economic side fire is equally influential. The probability, or often the practical certainty, of fire after the first cut, commonly determines lumbermen to leave no merchantable tree standing. Forest fires are thus the most effective barriers to the introduction of forestry.

Excessive taxation of timber land is another of almost equal effect. Because of it lumbermen hasten to cut and afterwards often to abandon, lands which they cannot afford to hold. This evil, which only the progress of public sentiment can control, is especially prevalent in certain portions of the white pine belt.

Forest Associations

Public sentiment in favour of the protection of forests is now widespread and increasingly effective throughout the United States. As the general understanding of the objects and methods of forestry becomes clearer, the tendency, formerly very marked, to confound ornamental tree planting and botanical matters with forestry proper is rapidly growing less. At the same time, the number and activity of associations dealing with forest matters is increasing with notable rapidity. There are now about thirty such associations in the United States. One of these, the Society of American Foresters, is composed exclusively of professional foresters. The American Forestry Association is the oldest and largest.

It has been influential in preparing the ground work of popular interest in forestry and especially in advocating and securing the adoption of the federal forest reservation policy, the most important step yet taken by the national government. It publishes as its organ a monthly magazine called *Forestry and Irrigation.* The Pennsylvania Forestry Association has been instrumental in placing that state in the forefront of forest progress. Its organ is a bi-monthly publication called *Forest Leaves.*

Other states which have associations or societies of special influence in forest matters are California, Massachusetts, Minnesota, Colorado, New Hampshire, Georgia and Oregon. Arbor Day, instituted in Nebraska in 1872 as a day for shade-tree planting by farmers who had settled on the treeless prairies, has been taken up as a means of interesting school children in the planting of trees and has spread until it is now observed in every state and territory. It continues to serve an admirable purpose.

Lumbering

All the operations of the lumber trade in the United States are controlled and to no small degree determined, by the peculiar unit of measure which has been adopted. This unit, the boardfoot, is generally defined as a board one foot long, one foot wide and one inch thick, but in reality it is equivalent to 14 4 cub. in. of manufactured lumber in any form. To purchase logs by this measure one must first know about what each log will yield in one-inch boards.

For this purpose a scale or table is used, which gives the contents of logs of various diameters and lengths in board feet. Under such a standard the purchaser pays for nothing but the saleable lumber in each log, the inevitable waste in slabs and sawdust costing him nothing.

In addition to this amount, an immense quantity of wood is used each year for fuel, posts and other domestic purposes and the total annual consumption is not less than 20 billion cub. ft. The years 1890 to 1906 were marked by rapid changes in the rank of the important timber trees with reference to the amount of timber cut and a shifting of the important centres of production.

Among coniferous trees, white pine has yielded successively to yellow pine and Douglas fir, while the scene of greatest activity has shifted from the Northern forest to the. Southern and from there is rapidly shifting to the Pacific Coast. The total cut of coniferous lumber has increased steadily, but that of the hardwoods is falling off and in 1906 it was 15% less than in 1899, while inferior hardwoods are gradually assuming more and more importance and the scene of greatest activity has passed from the middle west to the south and the Appalachian region. *Conifers. The* coniferous supply of the country is derived from four forest regions:

- The Northern forest;
- The Southern forest;
- The Pacific Coast forest;
- The Rocky Mountain forest.

The Northern forest was long the chief source of the coniferous lumber production in the United States. The principal timber tree of this region is the white pine, usually known in Europe as the Weymouth pine. It has an average height when mature of i to ft., with a diameter a little less than 3 ft., but the virgin timber is approaching exhaustion. White pine was one of the first trees to be cut extensively in the United States and Maine, the pine tree state, was at first the centre of production. In 1851 the cut of white pine on the Penobscot river was 144 million ft., that of spruce 14 million and of hemlock 11 million.

Thirty years later the pine cut had sunk to 23 million, spruce had risen to 118 million and hemlock had passed pine by a million feet. Meanwhile, the centre of production had passed from the north woods to the Lake States and for many years this region was the scene of the most vigorous lumbering activity in the world. Woods waste includes tops, stumps, cull logs and butts, but does not include defective trees left or trees used for road purposes. Mill waste includes bark, kerf, slabs and edgings. The spruce (*Picea rubens*) is used chiefly for lumber, but it is in large and increasing demand in the manufacture of paper pulp. For the latter purpose hemlock, poplar and several other woods are also employed, but on a smaller scale. The total consumption of wood for paper in the United States for 1906 was 3,660,000 cords, of which 2,500,000 was spruce. Of this, however, 720,000 cords were imported from Canada.

- The chief product of the Southern forest is the yellow pine. This is the collective term for the longleaf, shortleaf, loblolly and Cuban pines. Of these the longleaf pine, called pitch-pine in Europe, is the most important. Its timber is probably superior in strength and durability to that of any other member of the genus *Pinus* and in addition to its value as a timber tree it is the source of naval stores in the United States. The average size of the mature longleaf pine is 90 ft. in height and 20 in. in diameter. Shortleaf and loblolly are other important members of this group. Their wood very closely resembles that of the longleaf pine and is often difficult to distinguish from it. The trees are also of about the same size and height. Loblolly is, however, of more rapid growth. The total cut of yellow pine in 1906 was 11,661,000,000 board ft.; it has perhaps not yet reached its maximum, but is certainly near it.

  Another important coniferous tree of the Southern forest is the bald cypress, which grows in the swamps. The cut in 1906 was 839,000,000 board ft., a gain of 69% over 1899.
- But the great supply of coniferous timber is now on the Pacific Coast. The Douglas fir, also known as Douglas spruce, red fir and Oregon pine, is the foremost tree in Oregon and Washington and the redwood in California. When mature the Douglas fir averages 200 ft. in height and 4 ft. in diameter and the redwood 225 ft. in height and 8 ft. in diameter. Other important trees of the Pacific Coast are sugar pine, western red cedar, western larch, Sitka spruce, western hemlock and western yellow pine.

  Logging on the Pacific Coast is characterized by the use of powerful machinery and by extreme skill in handling enormous weights. This is especially true in California, where the logs of redwood and of the big tree are often more than f ft. in diameter. Logging is usually done by wire cables operated by donkey-engines. The journey to the mill is usually by rail. The mills are often of great size, built on piles over tide water and so arranged that their product is delivered directly from the saws and dry kilns to vessels moored alongside. The products of the Pacific Coast forest make their way over land to the markets of the central and eastern states and into foreign markets. Among the lumber-producing states, Washington has in seven years jumped from fifth place to first and its output has increased from 1,428,000,000 board ft. in 1899 to 4,305,000,000 ft. in 1906. Oregon and California have increased their output from 734,000,000 each in 1899 to 1,605,000,000 and 1,349,000,000 ft. respectively in 1906. Of the total output of these three states (7,259,000,000 ft.) 4,880,000,000 ft. is Douglas fir and 660,000,000 redwood.

- The important lumber trees of the Rocky Mountain forest are the western yellow pine, the lodgepole pine, the Douglas fir and the Engelmann spruce. The Douglas fir, here extremely variable in size and value, reaches in this region average dimensions of perhaps 80 ft. in height by 2 ft. in diameter, the western yellow pine 90 ft. by 3 ft. and the Engelmann spruce 60 ft. by 2 ft. Mining, railroad and domestic uses chiefly absorb the annual timber product, which is considerable in quantity and of vast importance to the local population.

COMMUNITY FORESTRY

Community forestry is an evolving branch of forestry whereby the local community plays a significant role in forest management and land use decision making. It involves the participation and collaboration of various stakeholders including community, government and non-government organisations (NGO's). The level of involvement of each of these groups is dependent on the specific community forest project, the management system in use and the region. It gained prominence in the mid-1970s and examples of community forestry can now be seen in many countries including Nepal, Indonesia, Korea, Brazil, India and North America. Community forestry is a branch of forestry that deals with the communal management of forests for generating income from timber and non-timber forest products. It has been considered one of the most promising options of combining forest conservation with rural development and poverty reduction objectives. Community forestry is defined by the Food and Agricultural Organization of the United Nations as "any situation that intimately involves local people in forestry activity". Community forestry exists when the local community in an area plays a significant role in land use decision-making and when the community is satisfied with its involvement and benefits from the management of the surrounding forest and its resources.

Community forestry is first implemented through the establishment of a legal and institutional framework including the revision of legal norms and regulations for forest management, the development of National Forest Plans and the strengthening of decentralization processes to sub-national levels of government. The second principal line of action is the implementation of pilot projects to demonstrate the feasibility of the community forestry framework. However, a study by the Overseas Development Institute shows that the technical, managerial and financial requirements stipulated by the framework are often incompatible with local realities and interests. A successful legal and institutional framework will incorporate the strengthening of existing institutions and enable the dissemination of locally appropriate practices as well as the local capacity for regulation and control.

Community forestry first came to prominence in the mid-1970s and has continued to evolve over the last few decades in a growing number of countries.

The availability of forest resources are often greatly reduced for use by the local people due to increasing pressures to cultivate the land, reliance on the forest resources and are also affected by economic and political changes. The evolution of community forestry in Nepal dates back to the late 1970s and was first instilled as an attempt to improve the management of forest resources and address environmental issues that were of great concern with the countries failing centralized forest policy. Over the past two decades, community forestry has been applied successfully in many developing countries, with its main goal being the alleviation of poverty amongst local forest communities and forest conservation. More recently, community forestry has been implemented in developed countries and it has been successful in its aims of sustainable forest management and securing socio-economic benefits for local communities.

STAKEHOLDERS

There is a large variety of stakeholders involved when considering community forestry. Participation from the various levels of community, government and non-government organisations (NGO's) is essential in the project's success. Stakeholders of community forestry have a vested interest to establish sustainable practices, whether this be to develop and maintain a regular income, ensure that forests are sufficiently protected to ensure their longevity or to reduce illegal activities and manage the area in such a way to promote tourism and conservation. Despite significant development, continued improvement in the collaboration between local governments and forest communities seems to be a key point for better community forest management. A wide range of futures scenarios have been put up to help the environmental decision process.

Challenges

A study conducted in the Brazilian Amazon determined that there are a number of challenges that must be faced when developing a sustainable management strategy for community forestry. The impact each management challenge has on other obstacles. The model is segregated into two phases: the 'development phase' during which several enabling factors (land ownership, organisational capacity, technical knowledge and capital) are needed to obtain a legal management permit and secondly the 'operational phase' where factors (clandestine loggers, access to markets, infrastructure and managerial skills) influence the successfulness of the management programme.

Best Practices

For a conservation programme within a Community forest to succeed, community involvement is imperative. Governments with interest in forest conservation introduce state wide policies and legislations which have historically failed to deliver the desired outcomes such as in China, Nepal and Peru. Moreover no single stakeholder by itself can ensure the success of such a programme.

Nepal

Common land in Nepal is owned by the state which often does not have the capacity to monitor and manage the area. This often leads to the over use of the resources by the community due to lack of incentives. To overcome this, programmes involving community participation were introduced and 'Forest User Groups' (FUG) formed to manage the forests resources without giving them ownership of the land. This has resulted in better outcomes in the region.

Indonesia

Villagers from three communities in Bantaeng district in Indonesia, with the assistance of a funded project obtained a Forest management license and secured a 35 year lease on their local forest. With the incentive to preserve their source of income, the villagers have had a positive impact on forest conservation. This is now viewed as a reference model for the Ministry of Forestry's future projects.

Korea

Farmers enter into a profit sharing contract with the 'Village Forestry Association' (VFA) to form a cooperative which assists farmers with reforestation in keeping with the legislation. VFA, though loosely linked to the Office of Forestry enjoys a degree of autonomy facilitating community participation. This system demonstrates the desirable mix of top-down and bottom-up planning ensuring government control as well as effective reforestation through active community participation.

Management System

As it is unlikely for any two areas to have identical socio-economic and environmental conditions, there is no blueprint management plan that would provide a blanket solution to all the problems. Based on research over several years in Nepal, it was noted that to have and effective management system, we need to identify variables which would affect the success of the system and group them into the following five sets:

I. Attributes of the resource system,
II. Attributes of the user group,
III. Attributes of the governance system,
IV. Attributes relating to interactions between the user group and resource, and

V. Attributes relating to interactions between the governance system and the resource.

In some cases, it is unrealistic to expect progress in a community level management of forest activities, as often conflict arise with respect to land use and benefit sharing within the community. Such issues can be overcome by recognising that a community level of management may not be the most effective management technique. Instead, adopting the following approach will provide a solution -

I. Management by smaller work groups within the community, sharing common interest in the resources.
II. A clear management plan with specific benefit sharing arrangement within the work group.
III. Develop management systems which are within the expertise of the working group.

Faith communities are increasingly participating in efforts to promote ecological sustainability. Whereas the last 50–100 years has seen them lease out their territory to industry, they are beginning to reclaim and restore this land. Their recognition amongst local and national authorities have allowed community forestry schemes to develop and there have been several highly successful efforts across Cambodia, which have been extended to Vietnam and Laos. In September 2010, Buddhists monks were awarded the UNDP's Equator Prize for their ongoing conservation work. This comprises the establishment of tree nurseries, seedling distribution, ordination activities, composting schemes and a vegetable garden. They are using this work to promote sustainable living and conservation throughout three districts; the pilot project, launched in 1997 in Kratie consists of a six pagoda network that covers 1,461 hectares (3,610 acres). The site has become a valuable source of information and environmental education as well as a base for the local community. Villagers regularly hold panel discussions about how they can go on to best take advantage of what the natural forest has to offer without destroying it.

Farm Forest for Household Use

I. Preserving the source of community's livelihood
II. Maintaining soil integrity by preventing erosion.
III. Use of fuel wood and fodder- Planting trees to provide fuel wood as well as fodder for cattle can be encouraged in ways that do not conflict with cash crops and food production. *e.g.* The Neem tree was introduced in West Africa and is now the most widely grown tree in the drier parts of the continent. It was easily cultivated and provides the farmers with good timber, fuel and shade.
IV. Developing nursery networks through support activities which assist with subsidies. This encourages entrepreneurship to produce tree planting stock for sale.

Farm Forests for Markets

Tree growing can take on the characteristics of a crop where there is a market for wood products such as poles, fuel wood and, pulp for production of paper. Companies tie up with farmers for supply of these products giving a steady source of income to the farming community. For example in Philippines, over 3000 farmers cultivate trees for pulp production for an industry that provides a market as well as a minimum price for the product, and Cooperatives of Village Forestry Association in Korea have helped local communities cater to a growing market in forest products such as timber and mushrooms.

ACHIEVE SUSTAINABLE FOREST MANAGEMENT

Historically, Indian forest policies have alienated people from the forests, thereby, exacerbating the rates of deforestation. Post-independence forest policies contributed to an expansion in agricultural production, met industrial demand for raw materials, and tightened control of forest lands through restricted access to forests and forest products. Protection policies increased the hardships of vulnerable social groups by denying them access to forests.

While the state took responsibility for managing forest resources, it did not have the commensurate resources to effectively manage and police the forests from traditional users. Before state intervention, forests were managed as communal property; the crucial role of forests in the economic subsistence of individuals, families and community was the basis for managing them as communal resources. A failure to recognize community control of forests led to a collapse in institutional norms that were instrumental in protecting and managing forest resources for local use. A shift in property rights to the state steadily undermined the rights of tribals to use and extract forest resources.

Involvement of rural communities living close to forests in protection and management of forest resources is enshrined in the National Forest Policy 1988.Translation of policy found expression in the resolution of Government of India, Ministry of Environment and Forests issued in June 1990.It envisaged that in lieu of the participation, the local communities will be entitled to sharing of usufructs in a manner specified by the concerned State Forest Departments. This led to the initiation of Joint Forest Management (JFM) programme. Importance of the programme is evident from the fact that the Government of India has constituted a "JFM Network" with the Inspector General of Forests, Government of India as the Chairman. The objectives of the network are (i) to act as a regular mechanism of consultation between various agencies engaged in JFM work in the country and (ii)To obtain constant feedback from various stakeholders on the JFM programme for proper policy formulation and suitable direction to States.

World leaders adopted the Millennium Declaration at the Millennium Summit in September 2000. The proportion of land area covered by forest globally is one of the indicators for the seventh MDG i.e to ensure environmental sustainability.In addition to quantitative, time-bound targets, the Millennium Declaration calls for other actions, including intensified efforts for "the management, conservation and sustainable development of all types of forests", an international commitment to sustainable forest management made in 1992 at the United Nations Conference on Environment and Development (UNCED) and embodied in the Forest Principles. Subsequent intergovernmental deliberations to promote progress towards sustainable forest management took place in the Intergovernmental Panel on Forests (IPF) and Intergovernmental Forum on Forests (IFF) from 1995 to 2000, and continue in the United Nations Forum on Forests (UNFF) as well as in other fora.

SOCIAL FORESTRY IN INDIA

Social forestry means the management and protection of forests and afforestation on barren lands with the purpose of helping in the environmental, social and rural development.

The term, social forestry, was first used in India in 1976 by The National Commission on Agriculture, Government of India. It was then that India embarked upon a social forestry project with the aim of taking the pressure off currently existing forests by planting trees on all *unused and fallow land.*

SOCIAL FORESTRY PROGRAMME

Government forest areas that are close to human settlement and have been degraded over the years due to human activities needed to be afforested. Trees were to be planted in and around agricultural fields. Plantation of trees along railway lines and roadsides, and river and canal banks were carried out. They were planted in village common land, government wasteland, andPanchayat land.

Involvement of Common People

Social forestry also aims at raising plantations by the common man so as to meet the growing demand for timber, fuel wood, fodder, etc., thereby reducing the pressure on the traditional forest area. This concept of village forests to meet the needs of the rural people is not new. It has existed through the centuries all over the country but it was now given a new character.

With the introduction of this scheme the government formally recognised the local communities' rights to forest resources, and is now encouraging rural participation in the management of natural resources. Through the social forestry scheme, the government has involved community participation, as part of a drive towards afforestation, and rehabilitating the degraded forest and common lands.

Need of Social Forestry

This need for a social forestry scheme was felt as India has a dominant rural population that still depends largely on fuelwood and other biomass for their cooking and heating. This demand for fuel wood will not come down but the area under forest will reduce further due to the growing population and increasing human activities. Yet the government managed the projects for five years then gave them over to the village panchayats (village council) to manage for themselves and generate products or revenue as they saw fit.

Types

Social forestry scheme can be categorized into groups; farm forestry, community forestry, extension forestry and agroforestry.

Farm Forestry

At present in almost all the countries where social forestry programmes have been taken up, both commercial and non commercial farm forestry is being promoted in one form or the other. Individual farmers are being encouraged to plant trees on their own farmland to meet the domestic needs of the family. In many areas this tradition of growing trees on the farmland already exists. Non-commercial farm forestry is the main thrust of most of the social forestry projects in the country today. It is not always necessary that the farmer grows trees for fuelwood, but very often they are interested in growing trees without any economic motive. They may want it to provide shade for the agricultural crops; as wind shelters; soil conservation or to use wasteland. Farm Forestry is another name for Agroforestry; a part of Social Forestry.

Community Forestry

Another scheme taken up under the social forestry programme, is the raising of trees on community land and not on private land as in farm forestry. All these programmes aim to provide for the entire community and not for any individual. The government has the responsibility of providing seedlings, fertilizer but the community has to take responsibility of protecting the trees. Some communities manage the plantations sensibly and in a sustainable manner so that the village continues to benefit. Some others took advantage and sold the timber for a short-term individual profit. Common land being everyone's land is very easy to exploit. Over the last 20 years, large-scale planting of Eucalyptus, as a fast growing exotic, has occurred in India, making it a part of the drive to reforest the subcontinent, and create an adequate supply of timber for rural communities under the augur of 'social forestry'.

Extension Forestry

Planting of trees on the sides of roads, canals and railways, along with planting on wastelands is known as 'extension' forestry, increasing the boundaries of forests. Under this project there has been creation of wood lots in the village common lands, government wastelands and Panchayat lands.

Schemes for afforesting the degraded government forests that are close to villages are being carried out all over the country.

Agroforestry

In agroforestry, silvicultural practices are combined with agricultural crops like leguminous crop, along with orchard farming and live stock ranching on the same piece of land. In lay man language agroforestry could be understood as growing of forest tree along with agriculture crop on the same piece of land.

In a more scientific way agroforestry may be defined as a sustainable land use system that maintains or increases the total yield by combing food crop together with forest tree and live stock ranching on the same unit of land, using management practices that takes care of the social and culture characteristic of the local people and the economic and ecological condition of the local area.

FORESTS AND THE MILLENNIUM DEVELOPMENT GOALS

The MDGs call for the integration of the principles of sustainable development into environmental policies. Environmental sustainability is being mainstreamed in forest policies around the world, particularly since UNCED, while the integration of the goals of poverty and hunger reduction in forest policies and plans is less widespread.

Community-based forestry, or participatory forestry, is particularly well placed to address poverty reduction. Community-based forestry is now well accepted and established in various countries in all regions, and programmes are beginning to generate financial and other benefits. Improving local peoples' rights and access to forest resources is a tool to the success of community-based forestry programmes. However, much still remains to be done to clarify and secure access rights. Many countries are working to strengthen forest governance, some through decentralization processes that allow the poor to derive more benefits from forests and be more involved in decision-making and forest management itself.

Although improving rights and access to forest resources and developing small-holder forest-based enterprises (including through community-private sector partnerships) show particular promise for poverty reduction, local political and economic realities, opportunity costs for the use of local resources, and other factors may prevent the poor from benefiting from community-based forestry programmes to the extent intended.

Intersectoral coordination is important for the achievement of all MDGs, but is particularly critical for reducing poverty and hunger and ensuring environmental sustainability, which are highly cross-sectoral by nature. Improved intersectoral cooperation and coordination will help efforts both to integrate the principles of sustainable development into forest-related policies and to integrate forests into sustainable development plans.

Joint Forest Management

Forest-based poverty reduction efforts tend to be linked to other land uses and should form a part of rural development strategies. Conversely, the potential for forests and trees outside forests to contribute to environmental sustainability cannot be fully realized without intersectoral cooperation and coordination. Intersectoral coordination, although difficult and time consuming, is necessary for sound decisions on land use and resource allocation, particularly when there are trade-offs between national development goals.

National Forest Policy in India treats forests as environmental and social resource. With the initiative of assigning ownership of Non Timber forest Produce (NTFP) to the local communities including the grass root level democratic institution for enhancing their livelihood opportunities and also improving their income with the value addition.

India has shifted the approach of forest management from regulatory to participatory mode of management with the resolution promulgated in 1990. At present, more than 17 million forests is managed by almost 10,000 Joint Forest Management Committees with the benefit sharing mechanism. In addition, the Government of India is in process to frame legislation for the settlement of tenurial rights of the forest dwelling communities mainly tribal on forests. This would definitely help in reducing the poverty of forest dwelling communities. The JFM resolution was circulated by Ministry of Environment and Forest in the year 1990 and 2000. JFM is a government resolution. A government resolution is a executive order or opinion of the legislature. A resolution does not have any legal backing.JFM as the term indicates is the management of forest by more than one party. In India there are two parties: the government represented by the Forest Department and the people living in villages located within forest or on the fringes.

There are two major reasons behind introducing JFM: one that the government's management system was not succeeding in arresting growth of forest degradation and deforestation.Second a new management paradigm was evolving in which the local people's participation was found to be an appropriate and promising tool in arresting forest degradation.

However in pre-independent India the concept of JFM didn't exist. The first National Forest Policy was adopted in 1894. Following were the guide lines

- Ensure maintaneance of adequate forest cover
- Meet the needs of local people.
- Collect maximum revenue after meeting the needs of the local people.
- Give priority to permanent cultivation over forestry land

In post independent India there was a shift in policy. In1988, the new forest policy was adopted which covered all the sustainable management approaches. The new policy had a few unique features. Which were as follows;

- Maintenance of environmental stability and restoration of ecological balance, soil and water conservation
- Conservation of natural heritage and genetic resources.
- Increasing productivity to meet the local needs then the national need
- Creating massive peoples participation movement to protect forest and tree cover and achieve the objective of reducing pressure on existing forests and meeting peoples need.
- Deriving economic benefits must be subordinated to these principal aims.

This initiated a process of reform at the local policy and operational level of forest management ensuring that the Forest Department developed close collaboration for protection and sustainable management of forests.

The aim was –Involvement of village communities and voluntary agencies of degraded forest land.

Important guidelines were as follows:

- The programme should be implemented under an arrangement between a voluntary agency or beneficiaries and the State Department.
- No ownership rights or lease should be given over the forest land.
- The beneficiaries should be entitled to share usufructs to the extent and subject to conditions prescribed by the State Government.
- Access to forest land usufructs should be available only to benefactress who get organized into a village institution especially for forest regeneration and protection. This could be through a village panchayat or a Village Forest Committee.
- The beneficiaries should be given usufructs like grass, lops and tops of branches and minor forest produce. If they successfully protect the forest they will be a portion from the sale proceeds when they mature.
- Areas selected from the programme should be free from claims from any person who is both a beneficiary under the scheme.
- The selected site should be worked in accordance of Working Scheme duly approved by the state government.Such a scheme may remain in operation for ten years and revised after that. The working scheme is prepared in consultation of with the beneficiaries
- It should ensure that there is no grazing at al on the forest land protected by the Village Forest Committee. Permission to cut and carry grass free of cost should be given so that stall feeding is promoted.

- No agriculture should be promoted on the forest land.
- Cutting of tress should not be permitted before they are ripe for harvesting. The Forest Department should not be permitted to cut to cut trees protected by the Village Forest Committee except in a manner prescribed in the working scheme. In case of emergency needs the village community should be taken into confidence.
- The Forest Department should closely supervise the work.If beneficiaries are unable to perform their assigned duties in a satisfactory manner the usufctory benefits will be withdrawn without giving any compensation.

Such set-up, however suffered from certain flaws:

- Bye- laws have not been formulated for the functioning of the JFMC though now most of the states have issued executive orders for the functioning of JFMC but these executive orders are not binding on JFMC. The aim was to decentralize the process and make JFMC.
- Minor Forest Produce (MFP) has not been defined neither by the State legislature or by the Centre.
- Central and State/UT Governments have issued guidelines for the creation and functioning of JFMC but these guidelines are not in conformity with the provisions of the Constitution
- Through the Constitution (Seventy-third Amendment) Act, 1992 and Panchayat (Extension to the Scheduled Areas) Act, 1996 ownership rights over minor forest produce (MFP) have been given to Village Panchayats. Now on the same resource base *i.e.* NTFP/MFP we have two sets of groups having ownership and while the Panchayats have a legal backing, JFMC don't have. conservation have been raised while the proponents of Panchayats have termed JFM as a parallel institution against the spirit of the Constitution.
- The JFM resolution only provides 20% of share to be given the Joint Forest Management committees, while rest of the income would go to the Forest Department. There is an obvious unequal distribution of benefit sharing between the parties. These committees would have to protect the forest for ten years,and would only receive 20% of the share.
- The national resolution provides that at least 33% of the seats shall be reserved for women in the specified committees. However states like Rajasthan are not following this provision.

Panchayati Raj System

The concept regarding Panchayati is contained in the Part IX of the 73rd amendment of The Indian Constitution.This amendment came into force in on 24th April 1993. It institutionalized the third stratum of government

(Panchayats) at the local level. Through the Constitution (Seventy-third Amendment) Act, 1992) Part IX "The Panchayats" was inserted in the Constitution which paved the way for "Village Panchayats" by making provisions for the constitution of Panchayats, their composition, election, powers, authority, responsibility, audit, etc. This Act doesn't apply to Schedule Area referred to in clause (1) and tribal areas referred to in clause (2) of Article 244 and certain other specified areas Village Panchayats have been given the responsibility of social/farm forestry, minor forest produce (MFP), and soil conservation through Eleventh Schedule.

"Panchayat" means an institution (by whatever name called) of self-government constituted under article 243B, for the rural areas. The amendment it fails to define what is 'self government'. It does not clarify whether self government means complete autonomously or extension of the state.It is left to states to derive their own interpretation regarding the nature of self government at the local level.By the word "nature" of self government it is meant reservation to be provided for the marginalized section of the society, the powers given to each level of self governing relating to administration, execution and performing the defined roles would be different in each State. Thus this will defeat the purpose of uniformity regarding the nature and concept of self government at an all India level.

The flaw again lies that the amendment has failed to assert the nature of 'Power And Authority'.This task is left to the state legislatures.Again defeating the purpose of uniformity. The consequences will be such that either the states will deny power to the panchyats instead of becoming autonomous local governing units they will basically remain in bureaucratic control.The burecratic trend can be seen most of the states.It is seen in the State of Goa power and functions have not yet been transferred to the Panchayats. As per the law there should be continuous elections every five years.However the statistics says that this status is not achieved in most of the states.None of the states except WestBengal, Tripura and Rajasthan have not been able to hold continuous elections every five years. The judgment provided by the Supreme Court states that it is mandatory for every state to hold elections to Panchayat.

Giving powers to the third stratum of government for effective management of resources at the district,intermediate and village level. Parallel institutions have come about for management of the natural resources which

Participatory Approach

According to United Nations Development Programme's report 67.7 million people belonging to "Scheduled Tribes" in India are generally considered to be 'Adivasis', literally meaning 'indigenous people' or 'original inhabitants', though the term 'Scheduled Tribes' is an administrative term used for purposes of 'administering' certain specific constitutional privileges, protection and benefits for specific sections of peoples considered historically disadvantaged and 'backward'.

Out of the 5653 distinct communities in India, 635 are considered to be 'tribes' or 'Adivasis'. With the ST population making up 8.08% (as of 1991) of the total population of India, it is the nation with the highest concentration of 'indigenous peoples' in the world. 68 million tribals who inhabit forests and wild lands throughout India, Adivasis have evolved an intricate convivial-custodial mode of living.

The Beginning of Land alienation

Introduction of the alien concept of private property began with the Permanent Settlement of the British in 1793 and the establishment of the "Zamindari" system that conferred control over vast territories, including Adivasi territories, to designated feudal lords for the purpose of revenue collection by the British. This drastically commenced the forced restructuring of the relationship of Adivasis to their territories as well as the powe relationship between Adivasis and 'others'. The predominant external caste-based religion sanctioned and practiced a rigid and highly discriminatory hierarchical ordering with a strong cultural mooring.

After the transfer of power, the rulers of the Residency Areas signed the "Deed of Accession" on behalf of the ruled on exchange they were offered privy purse. No deed was however signed with most of the independent Adivasi states. They were assumed to have joined the Union. The government rode rough shod on independent Adivasi nations and they were merged with the Indian Union. This happened even by means of state violence as in the case of Adivasi uprising in the Nizam's State of Hyderabad and Nagalim. The Constitution of India, which came into existence on 26 January 1950, prohibits discrimination on grounds of religion, race, caste, sex or place of birth and it provides the right to equality, to freedom of religion and to culture and education. STs are supposedly addressed by as many as 209 Articles and 2 special schedules of the Constitution - Articles and special schedules which are protective and paternalistic. Article 341 and 342 provides for classification of Scheduled Castes (the untouchable lower castes) and STs, while Articles 330, 332 and 334 provides for reservation of seats in Parliament and Assemblies. For purposes of specific focus on the development of STs, the government has adopted a package of programmes, which is administered in specific geographical areas with considerable ST population, and it covers 69% of the tribal population.

COMMUNITY PROTECTION FOREST

What the communities claim with respect to community protection forest is at odds with what the Department of Forests recognise. In the case of Dhuseri the constitution states an area of 532.5 ha. under community forestry while the operational plan refers only to 160 ha, reflecting an ongoing dispute between the FUG and the DFO over the area to be managed. There are several others cases *e.g.* Chautari and Parijat FUG where there are discrepancies in the stated figures on area between the original constitution of the FUG, the area demarcated in the operational plan and the information recorded on the Department of Forestry database.

Whatever the disputes at play, it is clear there are already major inequities between FUGs over the resources that they potentially have access to. This does not appear to have been addressed in any way by the Department of Forests. For Srijana FUG with 11.3 ha of community forestry ara and 0.07 per member, the potential benefits are going to be substantially less than for those members of Dhuseri or Parijat FUGs.

DEFINING COMMUNITY

Area and location is one matter. Another consideration is that of the way in which 'community' or 'user' is defined. Indeed there is disjuncture between the use of the term 'community' in community forestry and the rapid skip into the term 'user group'. There are various dimensions to this. As the regulations put it (Ministry of Forest and Soils Conservation, 1995, p.10)

'The district forest officer shall have to take into account the distance between the Forest and the village and the wishes as well as the management capacity of local users'. How 'account' is to be taken is of course not specified, but by raising the issue of 'distance' it is clear that more distant 'users' are at a disadvantage with respect to potential membership than those who are close. Categories of users based on differential use of the forest by product and time appear not to be officially recognised. This has meant that, for example the Tharu (the original inhabitants of the Terai), who tend to live further south from the FUGs and who had historically made seasonal but than regular use of forest products (for house construction) have effectively been disenfranchised from access although informal arrangements with some FUG committees have been established.

Buffer Zone Management

The buffer zone management around the Royal Chitwan National Park under the Parks Department is clear – community is defined in terms of residence. This has had the effect of disbarring household from wards in Rajahar VDC outside the buffer zone who had traditionally used products within the park boundaries from any access to the products of the buffer zone. However Department of Forests regulations allow households from within the park buffer zone to use the community forests of Dhuseri, Chautari and Bartandi. This asymmetry in rights of access appears to be leading to restrictions on residents from the buffer zone having access to the community forestry outside the park boundaries.

The story is however even more complicated. Many households appear to hold multiple membership of FUGs – for example many of the Buddha Mawali FUG members also have membership in another FUG. It was not possible to assess which household these were but analysis of the households that were not members of the FUG – and most FUGs have households living around the forest who are not members (in Srijana over 40% of households are not FUG members) tended to show that these were the poorer households, often the original residents of the Terai. There are at least two reasons why such households are not members.

The first hinges around the definition of what constitutes a legitimate 'use' of the forest. This has been largely determined by Forest Regulations, and adopted by most of the FUG constitutions, which refer almost exclusively to a restricted list of timber products. This does not include for example the right to make charcoal or graze goats in the forest. Thus the *Lodh* who have often engaged in ironwork and are dependent on charcoal from the forest have officially lost access to this resource although whether this consistently happens in practice is less clear. Those households in Bartandi who had established an important income source through goat rearing had been using the forest area of Chautari and Bartandi before the FUGS were established. Once the FUGs were established they lost their grazing rights and had to dispose of part of their herds. They have shifted into boulder collection from the nearby river.

ACCESS TO FOREST RESOURCES

The divide in access to forest resources between those who live in the north of the Terai and those who live in the south is likely to become a major distributional and equity issue in the future. There are therefore a number of complex issues in relation to how communities are defined and the determination of what area and where community forestry may be established. It is unlikely that the inequalities that have now been established can be resolved through reallocation of resources and the only possible route is a fiscal one, whereby communities that have gained control of valuable resources are appropriately taxed and the distribution of VDC expenditure deployed to address the existing inequities between communities with and without community forest resources and between communities that do have community forest. As matters stand in Nepal, this is likely to be a long and difficult route.

Legal or Encroachment Rights

This issue clearly matters more in the Terai than the hills and in closely related to (b). One's status as a 'user' at least in the view of the Department of Forests, clearly depends as to whether you have legal rights to the land on which you are settled. In one case (in Devdaha) resistance by the District Forest Officer (DFO) to the establishment of a FUG was expressed in terms that it could not be done because it would give legal status to illegal encroachment. The committee of FUGs do not appear to have adopted such a restrictive approach, although it must be recognised that encroachers and landless may well be amongst the poorest of households and the most dependent on forest resources for income, most notably through the collection and sale of firewood.

The processes by which FUGs come to be formed and established indicate a wide range in approach and participative mechanisms; these may have causal effects on the ways in which FUGs operate and deliver benefits although this is difficult to determine.

There is a strong contrast in the way in which community forest users group were established in Devdaha with heavy involvement of the NGO WATCH in the process of group formation and consultative processes and that of the HJAB FUG in Harpur which was essentially set up by the District Forest Office. Whatever the participative processes in bringing a CF group into being, there are at least two bureaucratic hoops through which all potential CF groups must go – the preparation and drafting of a constitution and the preparation and approval of an operational plan. The influence of these on the nature of the FUG is unclear, but the requirement that these documentary processes should be gone through put the District Forest Office in a strong position to regulate or control if and how the group is established.

There is not space here for a detailed textual analysis of the constitutions of the registered FUGs but a number of general points can be made. First is that they tend to be formulaic and have often been copied from other established FUGs. In Harpur the original name of the FUG from which the constitution (Hariyali CF in Rupandehi) had not been removed of the document. Second and related, the content of the constitutions largely address functions and structures following the listing of matters given in the 1995 Forest Regulations. The main headings required by the Forest regulations for User group constitutions and selectively illustrates these with extracts from the Dhuseri.

The extracts from Dhuseri, which do not differ substantially from other forest user group constitutions, are clear with respect to the stated objectives of the user group – the scientific management of the forest is the first amongst these, with meeting the demand for forest products by users the second. Dhuseri has established three membership categories, which relates to the way in which benefits are distributed.

The rest of the constitution largely deals with rules, committee structure and responsibility. This includes a five-member board of directors that includes a Chief of Board, with four councillors each with responsibility for one of the divisions of protection, plantation, management and utilisation. In other words the constitution proposes a village level version of the Department of Forests. In the case of Dhuseri, the strictures on crimes and punishment are held over to the Operational plan rules.

The Forest regulations (Ministry of Forest and Soil Conservation, 1995) also establish what should be included within the workplan and the key headings are summarise.

These regulations have since been backed up by Guidelines for the inventory of Community Forests (Ministry of Forests and Soil Conservation, 2000). These guidelines, which it is claimed have been developed to assist users and District Forest field staff in assessing the condition of the forests, are classic forest inventory. They are concerned with sampling design, stratification, sampling intensity, plot size and number, plot-layout, data capture, growing stock culminating in the estimations of annual increment and allowable cut.

But the exercise of estimating the annual increment and allowable cut is a fiction since a 1999 government order forbids the cutting of green wood, even though a later government order of May 2000 requires the calculation of annual allowable cut to be based on detailed calculations and estimates of annual increment. More to the point, and as Dhital et al (2003) have recently pointed out even the Department of Forests has limited capacity to implement these guidelines so how users groups can be expected to apply them is unclear. They found that of the 7048 community forests that had been handed over only about 21% of these (1518) had actually had an inventory.

Managing the Forest

It is also evident from the details on the methods cited above that this information is simply not relevant or usable by those who are meant to be managing the forest – namely the FUGs. In short the requirement for an operational plan, and the stipulation that a new one needs to be approved every five years sets has, as again Dhital et al, ' created a significant delay in forest handover and the renewal of [operational plans]'

As matters stand at present given the requirement and design specifications for constitutions and operational plans, the scope for participatory processes and genuine authority sharing is very limited. These bureaucratic devices, in the name of scientific forestry, can only be seen as serious impediments to promoting livelihood opportunities.

Product & Protection or Livelihood Oriented

As will be clear from the discussion on the content of the operational plans and constitution, the plans and objectives of these community forests combine a mixture of product and protection objectives and do not systematically address livelihood needs or recognise employment or income generation objectives for different social groups. If anything with recent government orders, forest protection is taking precedence even over production objectives.

It could be argued that because the Department of Forests of disciplinary necessity takes a single sector view of planning and development, foresters cannot be expected to explore areas of convergence between forest management and other institutional management structures in communities. Such an approach cannot address the 'joined up livelihoods' of people, particularly poor people, and the trade-offs they make in the management of their own and communal resources.

The message from this chapter is clear: all the evidence on institutional processes over which the Department of Forests has jurisdiction points to attempts to gain greater control rather devolve authority for community forestry in the Terai. The higher value of the forests in the Terai means that there are greater stakes to play for and through various means – control over land areas (how much, where etc.) and procedural strategies (constitutions, operational plans) – attempts to set limits to community authority and governance outcomes have been made. However field level processes, involving both DFO-staff and communities and NGOs, are another level of engagement and it is to these that we now turn.

COLLABORATIVE MANAGEMENT OF COMMUNITY FORESTS

In response to a community forestry environment that is complex and rapidly changing, CIFOR began a multi-country research project in 2000, which aims to enhance forest management decision-making at the local level.

This research project, 'Adaptive and Collaborative Management of Community Forests' (ACM), explores the potential role of *collaboration* and *social learning* in forest management, including the role of criteria and indicators (C&I) as a tool within that process. Research hypotheses include that self- or collaborative monitoring systems can support communities in deepening their knowledge about local systems and impacts of management strategies, as well as creating and focusing dialogue between diverse stakeholders. The underlying assumption is that these changes can (in some conditions) support equity, effectiveness and adaptiveness in community forestry decision-making.

The objective of the current research at the meta level is to generate insights into three questions. Does *collaboration* among forest stakeholders, enhanced by conscious and deliberate *social learning* processes in forest management, lead both to improved human well-being and to the maintenance of forest cover and diversity? If this is so, under what conditions does it occur? And, what are the key strategies, approaches and tools to enable these processes? These research questions are rooted in the assumption that the challenge of incorporating multiple interests at multiple scales into participatory interventions has not yet been successfully met in NRM.

APPROACH

The current ACM research is rooted in a participatory action research (PAR) approach. In most communities involved, diverse local people and other relevant stakeholders jointly developed a set of agreed and easily understood C&I. The process provided an opportunity for communication and learning within and across the stakeholder groups, especially with regard to visions and goals. The C&I set also provided a framework for later monitoring and assessing of key factors and their direction of change. This monitoring process creates the opportunity to feed information and learning back into the community forest management system. It thus serves to guide future action, helping to increase the sustainability of the community's forest resources. It was initially the researchers who offered and provided the framework for the social learning process; since that time local users have begun to adapt and apply these processes themselves. The ultimate goal is to completely transfer these, including the necessary facilitation skills.

Community forestry systems are complex and dynamic settings with multiple stakeholders, overlapping and differing interests, capabilities, and a myriad of challenging livelihood activities and processes. In some countries, such as those in Zimbabwe, the action research focus on collaboration has included power relations and negotiations *between* local peoples and other stakeholders. In other sites, such as Nepal, the focus has been primarily on stakeholder relations and equity *within* the local forest user groups (FUGs). Researchers there have tried to understand the stakeholder diversity within the FUG – based on *overlapping* categories including gender, caste, ethnicity, wealth and geography – in terms of issues of equity, power and access to resources and decision-making.

The short-term outcomes of the self-monitoring processes and follow-up actions appear to be contributing to positive change in this area. In follow-up to the monitoring workshops, for example, some of the forest user groups are shifting their committee-based decision-making processes (which were generally dominated by the elite) towards hamlet and interest group-based processes, including building mechanisms for feedback to the committee. Especially given the linkages between hamlets, ethnicity and wealth in some of the FUGs, these changes have the potential to help address some long-standing local equity issues.

A PAR methodology in isolation would present challenges in terms of producing generalizable results (a CIFOR mandate); thus, to enable generalizability, the PAR is embedded in a larger multi-site framework of scientific analysis. Specifically, in all research sites, researchers have laid the foundations for the comparative research by conducting a series of background studies elaborating stakeholder relations, historical, biophysical and socioeconomic contexts and initial levels of adaptiveness and collaboration. The studies took a consultative form of participation, but allowed researchers the time to build relations and the groundwork with local stakeholders for the main PAR phase of the research. Additionally, processes and learning in all sites are regularly recorded by researchers in a framework and format that is comparable across countries and sites. Other methods are also being used to triangulate the results across sites, countries and regions, key among these being the use of multivariate analysis across all sites and the analysis of the outcomes of the participatory modelling in Zimbabwe, Indonesia and Cameroon. Whereas the multivariate analysis is expected to provide a quantified picture of key drivers

for the success or failure of adaptive collaborative management processes, the analysis of the simulation models is expected to provide insights into the causality of failure or success arising out of the structure and behaviour of these processes. Ultimately, these elements of the larger framework for analysis will enable greater depth of understanding within each site and highlight findings that emerge across varying community forestry conditions.

Reflections

This is an ambitious project with high local and research expectations, and as such it faces some significant challenges. At the meta level, two of the most critical challenges are those of working across so many diverse sites and countries, compounded by the limitation of a very tight three-year time frame. Key challenges to working at the community level include: complex and pervasive hierarchical local stakeholder relations; low social capital; unstable political climates; and geographical isolation. However, key strengths include that, on the whole, community stakeholders, district and national partners, and field researchers have a high level of commitment to exploring the process and seeking local benefits – both social and environmental. The PAR and collaborative approach to the research incurs time costs to researchers but is enabling lessons to be relatively rapidly shared and incorporated to the research as it progresses.

Past CIFOR C&I research fulfilled its intention of generating useful and valid insights for some national, regional and global stakeholders through traditional social and biophysical research. In that research context, relatively few benefits were intended to accrue, nor did accrue directly to the communities where the research took place. The current research is focused on community-level processes and makes a conscious effort to bring good science into a coherent, integrated framework with local learning and benefits. The outcomes are not yet assured – the approach is new and certainly bears some risk. But the indications so far are that, in the context of these issues at least, a synergy exists between functional and empowering participation that will be well worth the costs.

2

Forests and Community Development

Because of their large number and wide distribution, the forest and wood-using industries influence the economic and social life of a great many communities. This influence is most keenly felt in the smaller cities and towns, particularly in wellforested regions, where they often constitute the dominant interest. In such situations, the prosperity of the community rises and falls to a very considerable extent with the prosperity of the forest-based industries. These have a decidedly stabilizing influence wherever the forests are managed on a sustained-yield basis, with the average annual cut and average annual growth in balance. Ghost towns do not occur where sustained yields are obtained and where mill capacity is adjusted to the growing capacity of the forest.

One of the important aspects of any industry is its influence on other industries and on service activities. The forest and wood-using industries purchase large quantities of axes, saws, tractors, trucks, planers, lathes, digesters, paper machines and other items of equipment; and they sell lumber, veneer, plywood, boxes, pulp, paper and similar items to other industries for use in the manufacture and marketing of their products. Railroads in some regions depend for their very existence on the freight revenue derived first from bringing raw materials, equipment and general supplies to communities dominated by one or more wood-using industries and secondly from transporting the products of these industries from mill to market.

Industry in general also supports many services, such as those supplied by grocers, druggists, clothiers, doctors, lawyers, ministers and teachers. The wood-using industries do not differ in this respect from other industries, except that in many rural districts they provide the sole or the main reason for the existence of a community. Where this is the case, they create the demand for such services; and where other industries are also present, they intensify that demand.

One further item of community concern is the relation of the forest and its dependent industries to the small landowners and especially to the farmer. The existence of a nearby wood-using industry provides a market, which in turn should encourage thrifty management of the wood lot. The additional income thus available can have a stabilizing effect both on the farmer and on the community of which he is a part. Forestry is a key to permanence.

FORAGE AND WILDLIFE

Not all forest industries are based on the utilization of trees. In many regions forage produced in openings in the forest and sometimes under the trees themselves, provides feed for cattle, sheep and pigs supplementary to that which they obtain from pastures and the open range. Beef, mutton and pork thus become in part products of the forest, the use of which for grazing may furnish an important means of support for the livestock industry. Timber production and livestock production are, however, not always compatible and stock must be kept out of areas where grazing is injurious to the forest. Browsing and trampling are likely to be particularly harmful in hardwood forests, from which domestic livestock should ordinarily be excluded.

Many fur-bearing animals which spend at least a part of their lives in the forest provide the raw material for another important industry. Milady must have furs; and mink, otter, fisher, marten, weasel, beaver, fox, muskrat, rabbit and many others supply them for her. Their production can well be one of the objectives of multipurpose forest management.

The taking of other forms of wildlife by sportsmen and fishermen helps to provide a market for manufacturers of arms, ammunition and fishing gear. Users of the forest for recreational purposes step up business for makers of automobiles, cameras, clothing, insect-repellents and gadgets of all sorts; for keepers of service stations, hotels, motels and dude ranches; and for packers, guides and wranglers. This stimulation of economic activity may extend as far back as the iron mine, the petroleum well and the cotton field.

Recreation, Inspiration

The most important influence of the forest as a source of recreation is not indirectly through industry, but on the recreationist himself. Its chief significance lies in the enhancement of human values rather than monetary values. The forest provides not only recreation, but "re-creation."

In the hurly-burly of modern life, with its economic pressures and nervous tensions, people need occasionally to get back to nature. Where better than in the open and particularly in the forest, can they regain both physical and spiritual strength?

As civilization has advanced, man's relation to the forest has continually changed. To primitive man the primeval forest was generally a gloomy and forbidding place. He viewed it with mixed feelings of fear and superstition. Tree worship was common in many tribes. "The groves were God's first temples"; they were also the first temples of pagan gods.

Gradually, as tools were developed and new uses for wood were found and as the demand for land for agriculture increased, man pushed back the forest frontier. Domination of the forest by man was a slow and difficult process which developed qualities of courage, hardihood, resourcefulness and ruthlessness. The men who lived and worked in the forest and whose hero was Paul Bunyan, were no weaklings.

Today throughout much of the world the early relationship is reversed. Virgin forests, except in parts of South America, Africa and Asia, are a rarity. There is a shortage rather than a surplus of wood for industrial use. A rapidly increasing population is becoming more and more concentrated in urban centers. Even the lumberjack usually lives in town or in a comfortable camp quite unlike the crude shacks of earlier days. Forest areas of outstanding value for recreation are shrinking even as the need for recreation becomes more acute.

Different people seek re-creation of body, mind and soul in different ways. Some like to hunt and fish with gun and rod; others prefer to use field glasses and camera. It is now the wild beast that fears man, not man the wild beast. Some like to lose themselves on camping trips in an inaccessible wilderness; others are satisfied to enjoy the beauty of the forest from the seat of an automobile. Many wish enough “roughing it” to give them the illusion that they are emulating the prowess of their forebears. Some seek health, others inspiration. All are attracted to the forest by the opportunity to find rest or strife, relaxation or stimulation, in an environment that takes them out of the stifling routine of their daily lives.

From time immemorial the forest has influenced the work of architects, painters, poets, dramatists and preachers. It has molded the character not only of those who live and labour in it but of those who visit it only occasionally to play or to pray. In the realm of the human spirit its gifts are invaluable both figuratively and literally.As Robert Marshall expressed it: “The most important values of forest recreation are not susceptible of measurement in money terms.

They are concerned with such intangible considerations as inspiration, esthetic enjoyment and a gain in understanding. It is no more valid to rate them in terms of dollars and cents than it would be to rate the worth of a telephone pole in terms of the inspiration it gives.” In face of mounting pressure for more wood for homes, furniture, magazines, rayon clothing and other tangible goods, it is important in the management of the forest to remember that “man doth not live by bread only.”Forests are an inescapable geographic fact. Not only do they dominate the landscape throughout much of the world, but they influence man’s life in countless and often unsuspected ways.

They temper the climate; improve the soil; reduce erosion and floods; support a wide variety of important industries; provide opportunities for the profitable employment of land, labour and capital; stabilize communities; and offer a unique source of recreation and inspiration in close communion with nature. Their management in ways that will assure the perpetuation of these values is an urgent task, of supreme importance to the people of the globe.

MOBILIZING ECOAGRICULTURE: FROM COMMUNITY ACTION TO GLOBAL IMPACTS

Ecoagriculture landscape innovators often identify their major constraint to be institutional barriers rather than technical or even financial ones. Key institutional challenges include inadequate community-level organizations for ecoagriculture action, landscape-scale planning, policies at various levels and mechanisms to achieve equitable outcomes in ecoagriculture landscapes.

ORGANIZATION OF COMMUNITIES FOR ECOAGRICULTURE

A core feature of ecoagriculture landscapes is the role of resident local farming or pastoral communities as key stewards, decision makers and managers of biodiversity. Public agencies may operate forests and protected areas, but their viability and sustainability depend on the matrix of private land uses in the landscape. Economic and social incentives can motivate collective action of local communities. Hundreds of community-based organizations have been documented to mobilize or engage in landscape-scale ecoagriculture initiatives. The institutions leading these initiatives are 'hybrids' fusing conventional farmer cooperatives, rural development committees and community-based conservation organizations. In The Philippines, for example, local farmer-based Landcare groups are linked with conservation organizations, municipal governments and research organizations to revegetate hillsides, conserve biodiversity in populated PAs and improve water quality. An important implication is the central role of communities in biodiversity conservation, especially outside PAs. Conservation organizations need to embrace and reorient their role explicitly to support local community stewardship in ways that respect and realistically address the central role of agriculture and livelihoods in planning and implementation methodologies.

Landscape-scale planning and governance

To achieve objectives at the landscape scale requires a process of collective action to support producers and coordinate action among key stakeholders in the landscape, often across sectors with a historical legacy of distrust. Development or adaptation of institutions for engagement, coordination and governance of ecoagriculture become the critical challenges. Scaling up and sustaining ecoagriculture landscapes that involve multiple stakeholders requires a process, and usually an institution, that will enable multi-stakeholder assessment, planning, implementation and monitoring for adaptive management. Currently, ecoagriculture initiatives take numerous forms, mobilized by community organizations, public agencies, NGOs or national/international projects. Methodologies that have been developed to assist the planning and governance process include landscape 'visioning' and 'scenario-building' processes, participatory landscape modelling, community biodiversity assessments and guidelines for 'adaptive collaborative management'. Multi-stakeholder trust-building processes and negotiation platforms are being adapted to the specific context of agriculture–biodiversity conflict situations (*e.g.* Hemmati 2007). Diversity of approaches is expected and desirable, but more systematic and comparative evaluation of effectiveness in achieving sustainable processes and outcomes is lacking.

Policies that support ecoagriculture landscapes

Ecoagriculture innovators around the world highlight the need for a more supportive policy environment for ecoagriculture, or simply the removal of major policy barriers. Core policy needs, at local, national and international scales are: (i) compatibility and coordination of agricultural development and biodiversity conservation policies, (ii) environmental legislation that embraces the potentials and rights of farming communities as conservators of biodiversity, and (iii) the removal of public subsidies for agricultural systems and investments that harm biodiversity.

Consumers, policy makers and investors are beginning to focus on the link between agriculture and conservation, and responding with new demands on the agricultural system, through systems of voluntary certification, industry standards and government regulation (*e.g.* SCBD 2005, Decisions III/11, V/5 and VI/5). Ecosystem/landscape-scale programmes and projects are being initiated by government agencies and NGOs, often in multi-stakeholder partnerships, and financed through public budgets (*e.g.* in India and China) and international development loans (*e.g.* Fernandes 2004).

Initiatives to achieve these policy objectives include the Central America Presidents' joint commitment to promote environmentally friendly agriculture, the removal of agricultural subsidies in Australia, and the recommendations from Communities to the Millennium Summit. New political coalitions are being formed to promote integrated cross-sectoral policies, bringing in voices and sectors not traditionally involved in either agricultural or conservation policy, such as municipal governments (in the context of political decentralization), urban consumer groups, international financial organizations concerned with screening investments for environmental sustainability, parts of the food industry, public health advocates and 'good governance' movements seeking to reduce wasteful spending on subsidies. At the international policy level, ecoagriculture strategies are being integrated into the work programmes of the relevant international conventions. For example, the CBD has adopted a new biodiversity goal of 30 per cent of agricultural areas under biodiversity-friendly management by 2010 (CBD 2006), and will focus on agriculture in meetings during 2008, as will the Commission on Sustainable Development. Rules developing under the World Trade Organization will need to be carefully scrutinized to ensure that they do not disadvantage producers in ecoagriculture landscapes. Some countries, notably Australia, Brazil and India, have adopted legislation that explicitly recognizes the rights of indigenous and other local communities to manage and conserve forests and natural habitats. The Convention on Biological Diversity and other international bodies are beginning to focus on opportunities for community-led conservation, although many elements of the conservation community are still uncomfortable directly addressing and supporting agricultural development. Policy changes can enhance the financial viability of ecoagriculture by removing government subsidies and fiscal incentives for biodiversity-harming production systems, in particular subsidies for agrochemical inputs and water, rules for commodity payment support that limit crop rotations, subsidies that favour annual crops over perennials, and intensive livestock production systems over grazing systems, and tax incentives for vegetation clearing.

Achieving social equity

Efforts to maintain or promote ecoagriculture landscapes are often instigated, complicated or impeded by the serious social inequities characterizing rural regions in many parts of the world. Indigenous communities are documenting their effective ecoagriculture systems in their efforts to reclaim land rights from the state. Corporate agribusinesses are seeking to promote ecoagriculture to ensure their 'social license to operate'. Ecoagriculture strategies that coordinate the use and management of landscape resources can help resolve resource disputes between farmers and pastoralists. Community groups and advocates for the poor are promoting ecoagriculture policies as a way to protect and restore biodiversity and ecosystem services important to the poor.

Ecoagriculture approaches can create more 'space' for equitable outcomes by identifying synergies between local livelihood benefits and benefits for agricultural economies and biodiversity, and by justifying stronger rights for poor producers over natural resources. But ecoagriculture systems are both context specific and the result of negotiations among diverse actors. To achieve equitable outcomes will require that poor and disenfranchised groups within the landscape organize themselves for political strength, that they join coalitions with other stakeholders, and that they be supported strategically in their negotiations with more powerful groups.

Major gaps

While this survey of ecoagriculture innovators reports numerous promising institutional models—at community, landscape and policy levels—the conditions under which such innovations are most likely to emerge, or can be successfully applied, are poorly understood. Effective cross-sectoral political coalitions have seldom arisen to advocate for reconciling conflicting agriculture and environmental policies. Ecoagriculture strategies are not well integrated into public investment plans, including the Poverty Reduction Strategies of low-income countries and donor strategies designed to support the MDGs. Rural farming communities are largely unrepresented at most international environmental policy forums, and environmental interests generally are absent from farming organizations. Local organizations find it difficult to access the specialized knowledge generated by others and are poorly integrated into ecosystem or watershed planning and policy processes at local, national or international levels. Few conservation organizations have staff with agricultural expertise. Most international and national policy and legal frameworks separate action on agricultural productivity, ecosystem management and rural livelihoods, and policy-making institutions reflect this separation. Most policy makers are unfamiliar with the opportunities for ecoagriculture, or of alternative policies and laws that would enable ecoagriculture activities and outcomes. Mainstreaming ecoagriculture will require that strategically important institutions—responsible for policy, research, and markets—modify how they do business to embrace ecoagriculture vision and strategies.

LOCAL CAMPAIGNS TO BOOST FARM FORESTRY

Two Farm forestry Service Groups (FFASGs) were formed in Kooki and Kakuto counties, each comprising five individuals. These groups are responsible for disseminating information on the role of Farm forestry to the communities as well its contribution to biodiversity conservation. They will give advice on nursery bed construction and management, and silvicultural practices, among others. In addition, after contacting and talking to some members working with forestry and community projects, and sponsored by RSGF, we have agreed to form a Rufford Small Grants Beneficiaries network. This network will share knowledge about their works. We will be able to advise each other and learn from each other in terms of successes and difficulties. The aim of the network is to carry out exchange visits with beneficiary farmers, stakeholders and other interested communities to facilitate knowledge sharing. This spirit is hoped to continue even in our next projects. At the moment four individuals have subscribed to this and we hope to bring more on board.

Main partners/collaborators

- Busia Women producers Association (BUWOPA), Uganda
- Integrated Rural Community Development Initiative (IRUCODI): the NGO that I am affiliated to.
- Pro-Biodiversity Conservationists of Uganda (PROBICOU)
- Wildlife and Animal Resources Management Department (WARM), Makerere University, Kampala.

Other Partners/Collaborators

- Natural Enterprise Development (NED), Uganda
- Uganda Women in Development (UWD)

Over the years, we have been approached by some NGOs, and individuals interested in knowing more about the work I have been doing with the help of RSGF. Some were seeking consultancies while others wanted to have an experience of the work we are doing in Rakai. It is important to note that from all the associations, meeting, and phone calls, we as a team have management to establish links with the above bodies. These links are bound to stay active and from them we shall expand on the work we are doing to include a broader community, and further promote conservation issues countrywide, as well as region wide. A few individuals have been enlightened and are compelled to apply for funds to do related work in other areas of the country. Such spirit expressed by young conservationists is very promising.

Introducing Social Farm forestry Networks

Networks have been formed among some beneficiary farmers and other stakeholders in Kabonela ad Kakuto (Figure 3. More networks will be formed as he project continues in the next phase of the project so that there is communication amongst all RSGF beneficiary individuals. The Social Farm Forestry Networks have been formed by the help of the newly created FFASG and by the end of the project period in future; it is projected that the entire district will have a lot of knowledge about farm forestry and its roles in biodiversity conservation s well livelihood improvement.

TREE PLANTING

Rolling out tree planting to other areas was based on the shared knowledge values in biodiversity conservation and improving rural livelihoods. Tree nursery construction and planting of seeds was started in late April 2009.

Table. Overall Number and Species of Trees Planted

Tree species	Quantity				Comment
	Phase 1	Phase 2	Booster phase	Total	
Eucalyptus spp	2500	2000	3000	7500	Income generation, cheap to raise and rapidly re-grows after being cut down
Pinus spp	0	800	1500	2300	Grow rapidly and straight and good for storm formation
Orange trees	0	800	0	800	Carbon sequestration, and nutrition in form of fruits
Mango trees	0	0	1100	1100	Carbon sequestration, and nutrition in form of fruits
Ficus spp	0	0	1000	1000	Soil improvement tree species and provides good habitat for many birds
Albizia spp	0	0	1000	1000	For holding the soil in place
Acacia spp	300	200	200	700	Nitrogen fixing
Moringa Oleifera	200	200	0	400	Herbal medicine
Total	3000	4000	7800	14800	

The parishes/villages of Kakuto, Kabonela and Kooki were the main stations.Seven thousand eight hundred trees were planted in the booster phase of the project; an excess of 1800 trees from the 6000 that were originally planned for in the project. In total, 14,800 trees have been planted since the commencement of the project.

Meeting with Local Leaders and Provision of More Extension Services

It is a rule of thumb to meet with Local council chairpersons, opinion leaders and key informants before start of any grassroots project. As routine these meetings were carried out and the way forward was drawn. The local leaders were very appreciative of the work and commended RSGF for their immense concern about nature conservation in the sub-Saharan Africa.

Continued Training

Stakeholders were educated about the following and given practical hands on in some areas:

- Silvicultural practices
- Seedbed management
- Fertilisation
- Increasing patchiness
- Increasing microenvironment
- Secondary succession
- Vegetation climax
- General Biodiversity conservation

ROLE IN CLIMATE CHANGE

The development of markets for tradeable pollution permits in recent years have opened up a new source of funding for tree planting projects: carbon offsets. The creation of carbon offsets from tree planting projects hinges on the notion that trees help to mitigate climate change by sequestering carbon dioxide as they grow. However, the science linking trees and climate change is largely unsettled, and trees remain a controversial source of offsets.

Climate impacts

Climate scientists believe human-induced global deforestation is responsible for 18-25 per cent of global climate change. The United Nations, World Bank and other leading non-governmental organizations are encouraging tree planting to mitigate the effects of climate change.

Trees sequester carbon through photosynthesis, converting carbon dioxide and water into molecular dioxygen (O_2) and plant organic matter, such as carbohydrates (*e.g.*,cellulose). Hence, forests that grow in area or density and thus increase in organic biomass will reduce atmospheric CO_2 levels. (Carbon is released as CO_2 if a tree or its lumber burns or decays, but as long as the forest is able to grow back at the same rate as its biomass is lost due to oxidation of organic carbon, the net result is carbon neutral.) In their 2001 assessment, the IPCC estimated the potential of biological mitigation options (mainly tree planting) is on the order of 100 Gigatonnes of carbon (cumulative) by 2050, equivalent to about 10 per cent to 20 per cent of projected fossil fuel emissions during that period.

However, the global cooling effect of forests from carbon sequestration is not the only factor to be considered. For example, the planting of new forests may initially release some of the area's existing carbon stores into the atmosphere. Specifically, the conversion of peat bogs into oil palm plantations has made Indonesia the world's third largest producer of greenhouse gases.

Compared to less vegetated lands, forests affect climate in three main ways:

- Cooling the Earth by functioning as carbon sinks, and adding water vapor to the atmosphere and thereby increasing cloudiness.
- Warming the Earth by absorbing a high percentage of sunlight due to the low reflectivity of a forest's dark surfaces. This warming effect, or reduced albedo, is large where evergreen forests, which have very low reflectivity, shade snow cover, which is highly reflective.

To date, most tree planting offsets strategies have taken only the first effect into account. A study published in December 2005 combined all these effects and found thattropical forestation has a large net cooling effect, because of increased cloudiness and because of high tropical growth and carbon sequestration rates. Trees grow three times faster in the tropics than in temperate zones; each tree in the rainy tropics removes about 22 kilograms (50 pounds) of carbon dioxide from the atmosphere each year. However, this study found little to no net global cooling from tree planting in temperate climates, where warming due to sunlight absorption by trees counteracts the global cooling effect of carbon sequestration. Furthermore, this study confirmed earlier findings that reforestation of colder regions — where long periods of snow cover, evergreen trees, and slow sequestration rates prevail — probably results in global warming. According to Ken Caldeira, a study co-author from the Carnegie Institution for Science, "To plant forests outside of the tropics to mitigate climate change is a waste of time.".

His premise that grassland reflects more sun, keeping temperatures lower, is, however, applicable only in arid regions. A well-watered lawn, for example, is as green as a tree, but absorbs far less CO_2. Deciduous trees also have the advantage of providing shade in the summer and sunlight in the winter; so these trees, when planted close to houses, can be utilized to help increase energy efficiency of these houses. This study remains controversial and criticized for assuming dark coloured trees might replace the frozen, white tundra in the upper northern hemisphere. Regular tree planting projects typically take place on lands that are only slightly different in colour. The warming impact was also measured over hundreds of years, rather than a 30-70 year time horizon most climate experts believe we have to fix climate change.

Furthermore, the described warming effect (of temperate and boreal latitude forest) is only apparent once the trees have grown to create a dense 'close canopy', and it is at precisely this point that trees grown for offset purposes should be harvested and their absorbed carbon fixed for the long-term as timber.

Costs

While the benefits of tree planting are subject to debate, the costs are low compared to many other mitigation options. TheIPCC has concluded that "The mitigation costs through forestry can be quite modest (US$0.1–US$20/ metric ton carbon dioxide) in some tropical developing countries.... The costs of biological mitigation, therefore, are low compared to those of many other alternative measures". The cost effectiveness of tropical reforestation is due not only to growth rate, but also to farmers from tropical developing countries who voluntarily plant and nurture tree species which can improve the productivity of their lands. As little as US$90 will plant 900 trees, enough to annually remove as much carbon dioxide as is annually generated by the fossil-fuel usage of an average United States resident.

Types of trees planted

The type of tree planted may have great influence on the environmental outcomes. Planting the wrong kind of trees, such asmonocultures of eucalyptus where they are not native species, can devastate the lands of the local people. However, it is often much more profitable to outside interests to plant non-native fast-growing trees, such as eucalyptus or pine(*e.g.*, *Pinus radiata* or *Pinus caribaea*), even though the environmental and biodiversity benefits of such monoculture plantations are not comparable to native forest, and such offset projects are frequently objects of controversy (see below).

To promote the growth of native ecosystems, many environmentalists advocate only indigenous trees be planted. A practical solution is to plant tough, fast-growing native tree species which begin rebuilding the land. Planting non-invasive trees that assist in the natural return of indigenous species is called "assisted natural regeneration." There are many such species that can be planted, of which about 12 are in widespread use, such as *Leucaena leucocephala*.

GENERAL CONTRIBUTIONS BY FARM FORESTRY PROJECT IN RAKAI

- *Biodiversity:* A variety of new life forms (animals, insects, micro-organisms, plants) have been established in the project area since the commencement of the project.

- *Soil management:* The stakeholders were taught how to use land within its capability; *i.e.* employing systems, types and intensities of land use with minimum risk of damaging soil resources.
- *Water quality:* Reducing salinity and water logging: Trees can reduce water tables thereby contributing to reduction in water logging and dry land salinity problems in agriculture.
- *Carbon sequestration:* Farm forestry implemented in Rakai, will help to sequester carbon on agricultural land because it can capture and store significant amount of carbon while still leaving the bulk of the land in agricultural production.
- *Wind breaks:* Trees hat were planted along the farm as shelters belts are protecting stock and crop from the damaging effect of the wind.

Apiary

Honey bee production: Education on production - the created Farm Forestry Extension Agents educated the selected stakeholders on purchase of bee colonies, management and production. The beehives were constructed and bee colonies distributed thereafter (Figure 10). They were supplied to the farmers who previously participated in the first and second projects, and had evidence of grown trees and land for apiary set up.

Some general outcomes from the project after the three project phases:

- Five tree nursery centers have bee established (two in Kakuuto, one in Kabonela and another in Kooki)
- 14800 trees have been planted and are expected to complement on the native vegetation
- Income generation y the stakeholders is evident through sale of seedlings
- Adoption of the seedling transplantation is on the increase in the district and many are not catered for especially those that Cannot afford to y the seedlings
- Biodiversity conservation is being taken to another level. There has been attitude change among the stakeholders and more are willing to take part in exercise and initiatives leading to biodiversity conservation.
- Twelve new families/households have been incorporated into the project and will be followed in the next phase
- Some farmers recently harvested their oranges ad others will be harvesting oranges and mangos in the next nine months.

Discussions

From a global point of view,it is widely recognised that we need to find ways to use agricultural lands in ways that produce environmental and social outcomes as well as sustaining livilihood. Farm forestry is one of the best methods of producing such outcomes. The farm foresty designed in Sangobay, Rakai, combines forestry activities with crop and animal farrmng. In farm forestry, our focus has been producng valuabble returns from the trees. However, apart from the returns farm forestry has led to important social networks that have in turn led to better orgaanisation and increased dissemination of knowledge regading biodiversuity conservation. Other benefits that it has brought to the area, region and probably at a gobal scale is maintaining soil and water quality and contribution to carbon sequestation.

Women and the youth are showing more interest iin tree planting and forest conservvation as a whole. The youth movements are encouraging alternative sources for water and other natural resources to the communities and this helps in preventing the local people from constant use of the Lake Victoria waters in which they end up dumping a lot of wastes. Together with the women's tree planting movement, the capacities and capabilities of rural women and youth groups are developing to implement programmes aimed at wise resource use and therefore restoring and maintaining the environment, the natural resource base and supply of tree products.

Such interest in tree planting is the cornestone to increased biodiversity consrvation in Rakai and Uganda at large. It is well understood that native vegetaion, especially remnant vegetation, provids the best range of habitat to sustain and enhance terrestial biodiverity (Gurr *et al.,* 2009). Protecton and expansion of vegetation areas is a vital component of biodiversity conservation.

However, in primarry production areas, adding farm forestry to the mix of land uses can contribute to posititve biodverity outcomes. Plantation forests can buffer edges between native forests and non-forest lands and thereby improving connectivity between forest habitats. Farm forestry shelterbelts provide benefits to birds, bats and insect biodiversity (Gurr e al., 2009); this is evident n the Sango bay project areas wich is increasingly becoming a suitable habitat for more bird species, micro-organisms and insects. this area is going to habor more biodiversity as the project continues and as more trees are planted in vital areas. Lyon *et al.* (2009) also confirmed that there were more abundant vertebrate populations in native forests than there were in cleared farm lands. This means that re-plantng certain cleared areas will bring back verterbrate populations as there will be creation of more habitat.

Farm forestry can help and protect soil resoures by protecting soils from wind and water erosion, increasing soil organic matter, fixing of nitrogen and other nutrients into the soil through tree species acting as hosts for micro-organisms, increasing soil carbon levels and cycling nutrients through the soil. According to Udawatta et al. (2008), establishing farm forestry buffers on previously cultivated agricultral areas has signifcant effect on soil quality in less than ten years. Benefits to soil fertility and therefore tree and crop productiviy can be enhanced by mixing species, as has been done in Rakai district. In several of the plantations, leguminous acacia species were incorporated. Accacia *spp* mixed with eucalyptus has been shown to improve productivity of both species due to the soil nitrogen fixing capacity of bacteria which populates acacia root system (Foresster et al., 2006).

Through inculcating new ideas, it was overwhelming to observe the level of positive response from the communities. Most of these ideas have been adopted by a number of stakeholders and are now being put into practice. This is evidenced by the number of tree nurseries mushrooming through out the district. The sale of seedlings to commnities has become a business in Rakai. This is good as it is promoting afforestation indirectly. This is also clearly shown by an increse in the number of trees around homeseads. This is truelly a good answer to the widespread deforestaton and unsustainable agricultural pratices; which are a significant part of the growing problem contributing to about 20 per cent of the current greehhose emiission (Leach, 2008).

Socio-Economic Factors Affecting Farmers' Decisions to Adopt Farm forestry in Sango-bay, Rakai District Exponential population growth in Rakai district has led to an increased demand for tree products (timber, medicinal herbs, firewood, etc) in the district resulting in high deforestation. Because the level of enforcement of the laws regulating the use of forests and natural habitats in the area is weak, the local communities have been depending on trees from their farms, neighbouring areas and those areas in the wetland without replacement.

Our project has been educating local people on various silvicultural practices and the art of harvesting trees for home consumption without degrading the environment and destroying biodiversity. The scope of the work up this period has been various parishes of Kabira and Kyebe in Kooki sub-counties in Kakuuto and Kooki counties. From our research it has been noted that the art of planting and harvesting trees for home consumption in many of the parishes has not been adopted well to the extent of self reliance, indicating that more education and demonstrations should be performed to enlighten them more. From the start this project was concerned with education, demonstrations, campaigns, and research. The research component aimed at identifying the socio-economic variables that affect the farmers' decisions to adopt farm forestry in Rakai district.

The maximum likelihood analysis results showed a positive relationship between age and the decision to adopt farm forestry. This indicates that age influences the farmers' decision to adopt farm forestry. The age of the farmer affected the farmer's knowledge and the awareness of the activities in the surrounding environment among other farmers.

The results showed a non- significant positive relationship between gender and the decision to adopt Farm forestry showing that males are not necessarily better adopters than females. Gender is thus not a critical issue in a farmer's decision to adopt farm forestry. That means that both males and females can equally adopt Farm forestry if the right measures such as education and demonstrations are equally provided.

A significant difference was found between the level of literacy among adopters and non-adopters. That means that formal education is a vital aspect in the farmer's decision to adopt Farm forestry and the fact that literate farmers would be adopters.

There was also a significance difference between in the adoption between those that were previously given lessons on silvicultural practices and demonstrations by the project team and those that were not. That means that spreading out education about farm forestry, biodiversity conservation and various silvicultural practices in many new areas would influence the level of adoption positively.

Land size is an indicator of the available economic resources and the willingness to adopt new technologies. This often revolves around factors such as the risk, preference, capital constraints, labour requirement and the tenure arrangements. In agricultural zones, tree crops compete with cash crops with the latter being preferred. Farmers in high potential areas are unwilling to divert land available for food and cash crops to trees which do not generate an equally lucrative product.

Non-farm income incorporates income earned by the household from different sources other than the farm. It is apparent that a non-farm income source varies greatly. This included trade, employment, casual work, credit, relatives, friends and miscellaneous sources. That means that the off farm income earned by the farmer does not affect the farmers ability to adopt farm forestry. This is because its investment is low cost. Furthermore, disposable income is the income that is left to the household to spend after taxation. It encompasses money accrued from different sources and used as expenditure for the household and savings. From the research, the household's level of income is a pre-disposable factor. It is not critical in the decision-making framework. Statistically, the decision to adopt is not based on the income level. This is attributed to the fact that tree seedlings are cheap and in other instances the farmers are given the seedlings free by organizations trying to promote Farm forestry in the area, as is the case with Integrated Rural Community Development Initiative, in Rakai District.

URBAN FORESTRY DEVELOPMENT AND MANAGEMENT

The urban forest is all of the woody vegetation growing in an urban area, including trees, shrubs, and vines found along city streets, public parks and private property. The City is responsible for managing an urban forest that contains over 40,000 street trees and 6,600 park trees (mowed areas).

There are several policies, programmes and projects issued and implemented for the past several decades which is an indication of continuing concern on the deterioration of urban environment. It is noticeable that new policies and programmes evolve whenever there is a change in administration (a common phenomenon in the Philippines) indicative of lack of continuity of previous initiatives. The major policies and programmes related to urban forestry are chronologically listed below:

- PD 1153 of Pres. Marcos dated 1976 (Tree Planting Decree to support PROFEM)
  - Requires all able-bodied Filipinos 10 years old and above to plant a tree per month for 5 consecutive years.
  - Certificates of planting and survival— requirement for graduation from school, renewal of job appointment and business permit and approval of retirement from service.
  - "Halamanan ng Bayan" launched by MHS to support this programme. It required each city or municipality to put up a nursery, garden and park.
  - Repealed by EO 287 dated July 25, 1987 because of dictatorial provisions and harsh penalty.
- PD 953 of Pres. Marcos dated July 6, 1976 (Greening of Private Lands Including Residential Subdivisions)
  - Requires private landowners to plant trees extending at least 5 m on each side of the rivers/ creeks.
  - Developers or owners of residential subdivisions and commercial/ industrial lots to set aside 30% of total area as open spaces for parks and recreational areas.
  - Penalizes unauthorized cutting, destruction or injury inflicted on naturally-growing or planted trees or vegetations in any public places.
- LOI 1312 of Pres. Marcos dated April 23,1983 (Establishment and Development of Local Government Forest or Tree Parks Throughout the Philippines).
  - Requires each barangay, municipality or city to establish and maintain at least one forest or tree park of considerable size.
  - MNR (now DENR) to allocate public lands for this purpose and to provide technical assistance and seedlings needed.
  - MHS to ensure that establishment of forest or tree parks is included in the land use plan of each barangay, municipality or city.
  - MILG (now DILG) to appropriate funds and implement establishment and maintenance activities.
- Memo Order Nos. 198 and 199 of Pres. Aquino dated November 9,

1988 (Luntiang Kamaynilaan Programme (LKP)/ Hardin ng Bayan Programme).

- Issued to help insure healthy environment in Metro Manila (MM) and to serve as model programme for other cities/municipalities.
- Anchored on the "Hardin ng Bayan" concept wherein each city or municipality should have gardens or parks of their own, transforming MM into a garden metropolis with lush vegetations, cool and fresh air like the countryside.
- Objective- to plant 2 million trees in 2-3 years and achieve a desired 1:4 tree-man ratio.
- For efficient, effective coordinated implementation, an Inter-Agency Committee (IAC) was formed: Co-chair- DENR and MMA (now MMDA); members- DPWH, DOTC, Metro Police Force, DOT, OPS and PMS.

• Memo Cir. No. 5 of Pres. Ramos dated August 27, 1992 (Clean and Green Programme).
  - Similar to LKP (same IAC composition except MMDA as chair/ lead agency) but wider in scope (not only greening but also cleaning activities)
  - Objective- massive planting (0.5 million trees/year or 2.5 million trees in 5 years from 1993-1997) to achieve the ratio of one tree for every 4 persons.
  - Although focused in MM, CGP has nationwide coverage and encouraging cities and municipalities to join nationwide contest for cleanest and greenest city or town.

• EO No. 113 of Pres. Ramos dated July 22, 1993 (Multi-sectoral Tree Planting Activities in Support of ENR Programmes/ECOREV).
  - Scope / Objective – regreening and rehabilitation of all open and denuded lands of public domain, idle lands, private lands and other suitable areas (both urban and rural) including rehabilitations of coastal and marine areas.
  - DENR to identify, assess and designate suitable area for planting and management and to provide technical assistance to participating agencies.
  - LGUs implement the programme in their respective level and set up counterpart funds.
  - Private sector participation encouraged via MOA or other appropriate arrangements with DENR.

• EO No. 118 of Pres. Ramos dated August 12, 1993 (Mandating the active participation of all government agencies nationwide in urban greening through an Adopt-A-Street/Park Programme)
  - Objective – greening of streets and parks in urban centers.

  - Requires all government offices and government owned/controlled corporations to adopt a street or park in coordination with concerned LGUs, NGOs and private sector by planting appropriate species and maintaining them for at least 5 years using their own funds/resources. and other resources.
  - DENR to manage and coordinate the programme through a designated National Coordinator.
  - Project to be turned over to concerned LGU for maintenance and protection.
- DENR-DILG-DPWH-CSC Joint Memorandum Circular No. 1 dated December 17,1993 (Implementing Guidelines for EO 118-Adopt-A-Stree/Park Programme)
  - Described the roles of each participating agency and outlined the schemes in the identification, selection and adoption of a street or park to be developed.
  - DENR to provide assistance to "adopters" in selecting suitable streets or park sites, in providing necessary planting materials and in monitoring performance.
- OPLAN SAGIP PUNO Programme of FMS-NCR/DENR launched on June 5, 2000.
  - Conceived as a component of "Lets Go Green Programme" of former DENR Secretary Antonio Cerilles.
  - Application of appropriate silvicultural treatments (*e.g.* removal of nails, wires/cables, water sprouts; surgical treatment of injured stem or root) to prolong life span and promote good health and vigor of trees planted in parks and along thoroughfares and streets in MM.
  - Supplemented by public awareness campaign.
  - DENR enters into MOA with participating agencies (*e.g.* subdivision homeowners association, city/ municipal government, NGOs, etc.)
  - DENR's role — conduct inventory and assessment of damaged/ injured trees; undertake appropriate silvicultural treatments; conduct information dissemination and training on tree care and maintenance; provide technical assistance and planting materials to sustain the project.
  - LGU's role — provide tree care and maintenance crews to sustain the project; assist DENR in information dissemination on maintenance and protection of trees.
- Proclamation No. 396 of Pres. Arroyo dated June 2, 2003 (Enjoining the active participation of all government agencies including government-owned or controlled corporations, private sector, schools,

civil society and citizenry in tree planting activity and declaring June 25,2003 as Philippine Arbor Day).

- Objectives- to promote multi-sectoral participation in tree planting nationwide; to develop greater awareness on the importance of trees in environment, health and human life.
- Participating agencies, LGUs, schools, etc. to identify areas to be planted in coordination with agencies which have jurisdiction over such areas *e.g.* DENR in case of public lands, LGUs in areas within their jurisdiction, DND for military lands reservation, DOT for ecotourism areas, etc.
- DENR, LGUs and schools — to establish and maintain nurseries.
- Respective participating agency/ instrumentality — to maintain and protect the planted seedlings.
- DENR — to provide technical assistance to all participants.

In general, the following goals and objectives are common to the Urban Forestry (UF) policies and programmes described above :

- To provide /maintain green, clean and beautiful environment;
- To promote public awareness on the importance of trees (promote environmental consciousness);
- Enhance people's participation in the programme;
- Promote multi-sectoral collaboration, cooperation and support; and
- In the case of LKP and CGP, the specific objective is to attain a 1:4 tree to person ratio to sustain ecological balance.

As strategy to enhance successful implementation of the project, DENR is usually tasked to provide technical assistance in planting, site and species selection, and maintenance operations, including provision of the planting stocks. Understandably, the DENR is also looked up to as the lead agency when inter-agency collaboration is involved in the programme. On the other hand, the city, municipal and barangay governments, which have jurisdiction over the project site, are usually tasked to maintain and protect the tree parks established and streets planted. They are also required to provide counterpart funds and other resources needed for these projects. At the end of each programme, there seems to be no serious post – project accounting or evaluation of outputs and accomplishments, including evaluation of success and failures. This may be attributed to the fast rate of turn – over of urban forestry/greening programmes being implemented. Another reason maybe lack of manpower and resources to monitor all the projects. For instance, in the case of Metro Manila, the Urban Forestry and Law Enforcement Division Office of FMS – NCR/DENR only has a small unit (Cooperative Planting Unit) under the Urban Forestry Section which is tasked to do the monitoring activities. Needless to say, the synthesis of lessons learned is an important input for planning and formulation of new programmes (*i.e.* we do not have to " reinvent the wheel" so to speak).

DECENTRALIZATION OF FOREST MANAGEMENT

Although a majority of forests continue to be owned formally by government, the effectiveness of forest governance is increasingly independent of formal ownership. Since neo-liberal ideology in the 1980s and the emanation of the climate change challenges, evidence that the state is failing to effectively manage environmental resources has emerged. Under neo-liberal regimes in the developing countries, the role of the state has diminished and the market forces have increasingly taken over the dominant socio-economic role. Though the critiques of neo-liberal policies have maintained that market forces are not only inappropriate for sustaining the environment, but are in fact a major cause of environmental destruction. Hardin's tragedy of the common (1968) has shown that the people cannot be left to do as they wish with land or environmental resources. Thus, decentralization of management offers an alternative solution to forest governance.

The shifting of natural resource management responsibilities from central to state and local governments, where this is occurring, is usually a part of broader decentralization process. According to Rondinelli and cheema (1983), there are four distinct decentralization options: these are: (i) Privatization – the transfer of authority from the central government to non-governmental sectors otherwise known as market-based service provision, (ii) Delegation – centrally nominated local authority, (iii) Devolution – transfer of power to locally acceptable authority and (iv) Deconcentration – the redistribution of authority from the central government to field delegations of the central government. The major key to effective decentralization is increased broad-based participation in local-public decision making. In 2000, the World Bank report reveals that local government knows the needs and desires of their constituents better than the national government, while at the same time, it is easier to hold local leaders accountable. From the study of West African tropical forest, it is argued that the downwardly accountable and/or representative authorities with meaningful discretional powers are the basic institutional element of decentralization that should lead to efficiency, development and equity. This collaborates with the World Bank report in 2000 which says that decentralization should improve resource allocation, efficiency, accountability and equity "by linking the cost and benefit of local services more closely".

Many reasons point to the advocacy of decentralization of forest. (i) Integrated rural development projects often fail because they are top-down project that did not take local people's needs and desire into account. (ii) National government sometimes have legal authority over vast forest area that they cannot control, thus, many protected area project result in increased biodiversity loss and greater social conflict.

Within the sphere of forest management, as state earlier, the most effective option of decentralization is "devolution"-the transfer of power to locally accountable authority. However, apprehension about local governments is not unfounded. They are often short of resources, may be staffed by people with low education and are sometimes captured by local elites who promote clientelist relation rather than democratic participation. Enters and Anderson (1999) point that the result of community-based projects intended to reverse the problems of past central approaches to conservation and development have also been discouraging.

Broadly speaking, the goal of forest conservation has historically not been met when, in contrast with land use changes; driven by demand for food, fuel and profit. It is necessary to recognized and advocate for better forest governance more strongly given the importance of forest in meeting basic human needs in the future and maintaining ecosystem and biodiversity as well as addressing climate change mitigation and adaptation goal. Such advocacy must be coupled with financial incentives for government of developing countries and greater governance role for local government, civil society, private sector and NGOs on behalf of the "communities".

SUSTAINABLE CREDIT FOR RURAL DEVELOPMENT

Few rural development endeavors confront the issues and challenges of sustainability more directly than the provision of credit to agricultural and rural producers. Agriculture lending, particularly to small farmers, has a nearly three-decade history of pairing donor agencies with developing country governments to promote economic development by alleviating credit constraints. Despite the relatively large amounts of credit directed towards rural producers, results have often been disappointing.

Provision of agricultural credit has been pursued in the context of both production and equity objectives. The strategy generally employed is to increase the total supply and reduce the cost. By making cheap credit widely available to the rural poor it was thought that production would be increased, with the added benefit of improving the income distribution of the rural poor. Credit policies and programmes have been based on a set of "heroic" assumptions about small rural producers and their access to financial services. These include, for example, that small producers must be induced to adopt agricultural innovations by offers of cheap credit; that they have no savings capacity; and that with limited access to commercial credit they must rely on "usurious" informal sources, such as money lenders or pawnbrokers.

The lending landscape created by acting on these assumptions features convoluted regulations, special funds, complex rediscounting and reserve arrangements, loan guarantees, and other incentives to encourage lenders to increase credit flows to the rural poor. Special programmes and institutions, each with their own objectives, targeted particular groups to get credit, in what amount, and for which. Financial markets have become fragmented, rather than integrated, as new entities, such as agricultural credit agencies, development banks, and cooperatives were created to disburse and monitor these supervised credit programmes. With such a narrow focus on credit disbursement, many lenders either failed to, or were even prevented from providing deposit and savings services to rural producers.

Besides dysfunctional fragmentation, rural financial markets exhibit high degrees of inefficiency, ineffectiveness, and distortion. Populating the credit landscape are institutions with programmes plagued by loan recovery problems, outright failure to repay, chronic dependency on subsidies and external support, and excessive transactions costs for both the institutions and clients.

These institutions remain viable only due to continuous infusions of external resources. Further, they rarely reach their intended small farmer beneficiaries, and fail to achieve either their production goals, or any substantial increase in rural income.

Despite the history of failure, we cannot ignore the potential contribution that effective financial markets and financial institutions can make to economic development by stimulating savings and facilitating intermediation between borrowers and savers. There is a strong development rationale for public intervention to build up the formal (and informal) financial sector and extend services to groups that are not yet being adequately served. However, to be of ongoing benefit to developing countries, these institutions must prove sustainable over the long term.

Because the "good" that credit institutions deal with is monetized, relatively precise calculations of the cost and value of the benefits provided can be made, relationships between benefits and incentives can be quantified, and measures for sustainability developed. SCOPE highlights stakeholders and the value they attach to institutional outputs, factors that play key roles in rural credit. The framework also focuses attention on the complex and multiple exchanges between institutions and actors in the environment.

This chapter examines the credit component of the USAID-funded Provincial Area Development Programme (PDP) in Indonesia using the SCOPE concepts. Credit systems are reviewed in three provinces: Central Java, South Kalimantan, and West Nusa Tenggara.

The case study focuses on the development of capacity and performance of core insti tutions involved in the various PDP credit systems; the impact of the provincial credit systems on borrowers and other beneficiaries; and the credit systems' viability and prospects for sustainability. The study seeks to convey a flavour of the complex interactions among institutional actors that played an important role in shaping the development of the credit system.

THE PROVINCIAL AREA DEVELOPMENT PROGRAMME

Indonesia's PDP was implemented over a ten-year period, beginning in 1978 in two provinces, one of which was Central Java. It expanded to six, including South Kalimantan, in 1979, and encompassed eight by 1980, reaching West Nusa Tenggara in that year. The project design provided resources and support for the planning, implementation, monitoring, and evaluation of small subprojects, to increase the incomes of the rural poor by decentralizing development efforts to the provincial level and below. In the eight provinces, PDP undertook a wide range of sectoral activities, including agricultural food crops, estate crops, livestock, fisheries, village industry and handicrafts, manpower training, social welfare, and cooperative development.

PDP's objectives included improving

1. The capability of local governments to undertake rural development activities that improve the productive capacity of the rural poor;
2. The capability of the central government to support local governments in planning, implementing, and evaluating activities that improve the productive capacity of the rural poor; and
3. The incomes of the rural poor within the project areas through implementing small subproject activities. These objectives shifted over time, especially after 1982. They gradually moved away from experimental, direct, beneficiary-oriented subprojects, which were expected to be taken up by local government agencies, towards the development of institutional capability and introduction of planning and monitoring systems.

The project worked at the central level with the national planning agency, and at the provincial level, beginning in 1980, with decentralized regional planning boards (BAPPEDAs). PDP also worked through local sectoral agencies (*dinas*) at the provincial level and at the rural district (*kabupaten*) and sub-district (*kecamatan*) levels with local government entities.

From the outset, participating provinces were required to make some arrangement for the provision of credit in support of the various sectoral development efforts. How credit was structured and provided, varied considerably from province to province. Initially, the dinases handled the details of the credit programme through individual subprojects, but this proved to be a disaster. The local sectoral agencies had little difficulty in supplying credit as input to their production efforts, but had little capacity to manage repayment.

Demand for credit in PDP came from the sectoral agencies to support their production efforts in the many subprojects that were being sponsored. Stimulating commodity production was the main focus of effort, and getting funds out to producers in support of the subprojects was the major emphasis of credit. Since lending is only one element of the activities a viable credit system needs to engage in, financial services delivery to beneficiaries began to founder. Repayment, covering operating costs, and savings mobilization were all neglected.

One notable exception to the pattern of failure was the sub-district credit agency (badan kredit kecamatan), or BKK, in Central Java province, supervised by the Regional Development Bank (BPD) and supported by PDP. Here the emphasis was on institutional development over strict credit delivery. The success of the BPD's Central Java BKK operations during the early years of PDP attracted considerable attention, and led to emulation in other provinces. As a result, the BPD was assigned the responsibility for providing PDP credit, including credit management, training and supervision, through the BKKs' networks of local offices. In conformity with the PDP objective of targeting the rural poor, credit operations were directed at small enterprises that were not served by formal lending institutions.

The three provincial BKK systems reviewed subsequently, illustrate PDP efforts directed at creating and maintaining viable institutions capable of responding to the demand for financial services in rural Indonesia. The three credit systems vary with respect to size, scope of operations, facilities, staffing, management, and leadership. In part, this variation reflects different strategies being pursued in the delivery of financial services, but it is also a function of the different levels of development of these systems.

The Central Java BKK

In 1970, the BKK system was created as a project of the Central Java provincial government, which capitalized each individual local unit. By 1979, about one-third of the units had ceased to exist or were operating at such low levels as to make them non-viable. PDP provided support to rehabilitate the BKK. Interventions began in 65 units in the five rural sub-districts designated as PDP's target area. The institution-building experiment consisted of:

1. Recruitment and training or retraining of staff;
2. Development of accounting, supervision and auditing, reporting and classification procedures for the system as a whole; and
3. Increased capitalization on a graduating scale based upon unit classification.

The intent was to deal with immediate performance problems and build future capacity as well. Within the first three years of the experiment, substantial progress was made in improving the operations of targeted units. The success attracted the attention of the central government, which in 1981 provided a loan of $4.7 million to extend the rehabilitation efforts through the remainder of the province's sub-districts.

Size and Scope of Operations

The Central Java BKK has units in 497 sub-districts, and operates over 3,000 mobile posts that follow village market day schedules. It serves around half a million clients, and continues to grow. The BKK is primarily a lending institution, but targets savings mobilization as well. Compulsory savings is a loan requirement designed to inculcate new behaviours and to increase available loan capital. Following a change in regulations, the BKK now offers savings accounts to non-borrowers and is aggressively pursuing this new market. Average loan size is $36, with the borrowing period averaging 12 weeks.

Facilities and Staff

For the first ten years of its existence, the BKK operated out of borrowed rooms in the kecamatan government complex with borrowed or nonexistent equipment and material. Units were staffed with local government employees. The PDP institution-building intervention led to changes. The project established a revolving fund, managed by the BPD that BKK units meeting profitability criteria could borrow from, for equipment. Permanent offices were financed the same way, using funds from the provincial development budget. This practice limited public subsidies and put government resource allocation on a profitability basis.

In the mid- 1980s, the BKK shifted its staff from civil servant status to employees of the system itself, which allowed the introduction of performance-based pay, profit-sharing incentives, and more flexibility to hire and fire. Each unit has three core staff, a manager, cashier, and bookkeeper; hiring additional staff depends on service demand and profitability.

Leadership and Management

Key actors providing leadership for the BKK have been the director of the BPD, and an assistant to the provincial government secretary. They maintained linkages with the governor's office and the BAPPEDA, lobbying and negotiating for policies favourable to the BKK. The system also has a supervisory board that provides both leadership and oversight, and reports to the district head. The BKK's official designation as a sub-district institution has given the sub-district head a monitoring role, which has had a positive effect on performance. In addition, village chiefs serve as guarantors of the reliability of individual borrowers and as information conduits regarding credit matters.

The BKK in South Kalimantan

The BKK in South Kalimantan is not only a direct product of PDP credit interventions in the province, but as its name implies, is also a conscious replica of the Central Java system. Established in 1985, the South Kalimantan BKK has deliberately limited its services to the provision of small, short-term loans. The BKK serves 28 sub-districts of the 106 in the province. Loan packages in South Kalimantan tend to be larger and about twice as long in duration as those in Central Java. A decision not to offer a savings programme at the outset was based on considerations for the difficulty in training new staff to perform complex bookkeeping functions for two separate accounts simultaneously. The BPD has stated that it is making plans for the addition of a compulsory savings programme.

Facilities for the BKK, including equipment and motorcycles, were financed by the district with PDP funds, which are not subject to repayment by the system. PDP provided another subsidy in the form of two years' worth of staff and operating costs for each BKK unit. Staffs are employees of the credit agency, with three per unit.

As in Central Java, the BPD director has played a lead role in maintaining close contact with the provincial development planning board and the province governor on BKK matters. The system does not have a high-level supervisory board, though it does involve the sub-district head in performance monitoring.

The Lumbung Kredit Pedesaan in West Nusa Tenggara

In West Nusa Tenggara, the period 1980-86 saw PDP experimentation with lending through precooperatives (village training centres) managed by the provincial development planning board, later through mobile banks provided by the BPD, with a simultaneous cooperatives lending scheme, in which the BAPPEDA, the dinases, and the BPD all played a role. These experiments led to failure due to a variety of classic credit problems: loans to groups with no prior identity, loans for economically nonviable enterprises, inadequate lending and collection procedures, and poor interagency collaboration.

Despite resistance by the BAPPEDA, PDP established a new credit system in West Nusa Tenggara in 1986, the Lumbung Kredit Pedesaan (LKP), to turn the credit situation around. The LKP was the last of the PDP credit systems to be established. The system evolved towards the Central Java BKK model, kecamatan-based with mobile units for village outreach, although the LKP is conceived as more strongly linked to village locations. Average loan duration is 12 weeks, the average amount, just under $40. LKP loans include compulsory savings. A unique feature of the LKP is that in some locations, the system has been able to mobilize voluntary savings both from active borrowers (in excess of the compulsory requirements) and from nonborrowers. This is impressive since savings deposits do not bear interest.

LKP units finance rental of office space through profits, but PDP has given them equipment and motorcycles, and provided a two-year salary and operating expense subsidy, as in South Kalimantan. LKP units' core staff adds a messenger and night guard, bringing the total staff to five. Staffs are not civil servants, and village chiefs have a say in hiring. Leadership comes from the regional development bank, the provincial development planning board, kabupaten officials, and village chiefs. However, the sub-district head does not have a role.

Success And Sustainability PDP's final evaluation concludes that, "the credit programme was a stellar example of the manner in which a PDP initiative, once adequately tested, was recognized as worthy and was readily transferred, modified and adopted in other provinces". Indeed, the revitalization of the BKK in Central Java, and the later establishment of new credit systems in South Kalimantan and West Nusa Tenggara, demonstrate how a flexible and experimental approach to design and implementation can lead to success.

However, what is success for the credit component in terms of sustainability? As noted earlier, the objectives and the institutional approach to the provision of credit under PDP underwent a process of transition and evolution. At the outset, PDP implementors seldom saw the provision of credit as an end in itself, but rather as a necessary support element to the continued achievement of results by the sectoral agency commodity subprojects to which the provision of credit was subordinated. This approach proved to be nonviable financially, leading to the experimentation with the BKK model, which reversed the emphasis of the credit component, highlighting the viability of the credit institution rather than the outcomes of the sectoral subprojects. This amounted to a change in objectives.

By the late 1980s, the imminent termination of PDP funding provoked considerable discussion of a "break-even point" for both the individual BKK/LKP units and the provincial systems. Concern for institutional mechanisms that could guarantee financial viability and sustainability accelerated the move away from the rural producers in the sectoral subprojects, the original priority beneficiaries (who were high risk borrowers), to lower risk client groups. For example, the South Kalimantan BKK retreated from the seasonal agricultural loans that were once a major percentage of its portfolio, due to high levels of arrears and low turnover. This trend left key stakeholders with interests in PDP's development objectives dissatisfied. They argued that the shift to an institutional approach to credit abandoned the very sectoral agency commodity programmes that credit was originally intended to support.

This situation highlights a critical issue for sustainability. The SCOPE definition of sustainability refers to the ongoing production of goods and services sufficiently valued such that inputs continue to be provided. The "technology" of credit provision specifies what is needed to maintain this input-output relationship without continuous in injections of outside resources; for example, lending criteria that limit risk, savings mobilization, appropriate interest rates and rate spreads between loans and other accounts, administrative and supervisory cost containment, and so on. The issue arises when the question is asked, who values the goods and services and what are their interests? It poses problems for sustainability when the answer reveals conflicts between the valuation of the outputs and technical requirements for their sustained production. Indonesian government sectoral agencies valued the credit programmes as supports to the experimental commodity production activities, which involved greater risks for lending. As the BKK/LKP systems made changes to increase financial viability, and ultimate sustainability, they emphasized short term loans to low risk clients, such as traders and food processors, and raised interest rates in some cases. The sectoral agencies dealing with food crops, livestock, fisheries, and industry complained that their clients were not qualifying for loans and demanded preferential treatment.

Thus important stakeholders in the credit systems did not value the modified services the BKK/LKP was providing. The changes increased the prospects for the systems' financial viability, but in terms of the SCOPE definition, this lowered the prospects for sustainability. BKK/ LKP response was to argue that the shift away from the sector programmes would only be temporary. Once the credit systems were firmly established they could afford to carry a more risky portfolio and serve more of the clients of the sectoral agencies.

The stress on institutional development objectives for the credit component, besides changing the BKK/LKP product mix, introduced a new set of actors. The BPD became a major stakeholder and to some extent displaced the sectoral agencies and BAPPEDAs as supporters of the programme. The BKK/LKP systems are seen by some local government actors as a kind of institutional "imperialism" on the part of the BPD. Increasing emphasis on the expansion of financial services to include compulsory and voluntary savings is taken as further evidence that institutional viability has come to replace the original objectives of credit delivery. The PDP credit component case illustrates a situation where development and sustainability objectives appear to be in conflict. The institutional approach to the provision of credit by building up the BKK/ LKP pursued three objectives:

1. Guarantee access to credit for the rural poor, who were identified as the PDP target group;
2. Cover the operations and management costs of the BKK/LKP systems and their individual units;
3. Provide for measured growth of services and of the system to meet future needs and demands.

The first of these is a development aim, the other two relate more to the ongoing viability of the credit institutions. The actors involved had differing priorities among the objectives; USAID, the BAPPEDAs, the dinases, provincial governors, the BPD, and rural residents all had various preferences, interests, plus differential ability to make theirs prevail. To a large extent, the institutional sustainability targets conflicted with the development objective.

For example, one of the factors contributing to the failure of the earlier PDP credit schemes resulted from pursuing the development objective of targeting loans to groups participating in sectoral agency subprojects. Preferential treatment of a class of borrowers because of their involvement in a subproject led to misinterpretation of borrower obligations and high default rates. Without loan repayment, the credit agencies could not remain viable lenders.

Considering the time dimension heightens the potential for goal conflicts. What leads to financial viability in the short term, may inhibit achievement of sustainability and development objectives in the long term. For example, Dukesbury suggests that the Central Java BKK's success may be attributed in large part to its almost exclusive concentration on twelve-week loans to small traders. The BKK's record of high capital utilization, quick turnover, low default rates, and moderate increases in average loan size led the LKP to impose a similar restriction. Immediate institutional viability was enhanced, and it appears that the small entrepreneur market is far from saturated. However, over the long term, failure to expand into other sectors deprives those sectors of needed financial services and exposes the BKK/LKP to the potential danger of excessive reliance on undiversified portfolios.

In addition, all of the BKK/LKP systems have been and continue to be subsidized; for example, the BPD has undertaken most of the supervision of the system units. This and other subsidies have promoted short-term viability, but it remains difficult to determine how and when they may cease without damaging the systems. The Central Java BKK is in the best position, having fewer subsidies to start with and better possibilities for expansion. Without steady growth in the South Kalimantan and West Nusa Tenggara systems, erosion of performance is a risk. Eliminating subsidies and maintaining viability is likely to be a function of attaining, and then surpassing, a minimum size level.

USAID's final evaluation of PDP indicates that an effective credit delivery system has been created under the project, but as Dukes-bury notes it is uncertain whether the BKK/LKP can become institutionalized successfully in other Indonesian provinces beyond Central Java following the termination of outside funding. The systems in South Kalimantan and West Nusa Tenggara are still fragile. Thus BKK/LKP's long-term sustainability is open to question as the systems seek to expand in the future. The chapter's discussion detailed PDP's success in introducing, testing, and modifying institutional innovation in credit provision at the village level.

The Indonesian experience shows that a decentralized, flexible, consciously experimental approach to discovering what works is particularly well suited to local institutional development. Further, the case exemplifies the positive interaction between performance gains and stakeholder support for further investment; it was the BKK's visible, early success in demonstrating viability that led to decisions to pursue system development, both in other Central Java districts and in other provinces. The SCOPE model's definition of sustainability highlighted a conflict between the development objectives of PDP's credit component and the requirements for institutional viability and sustainability. Reaching the rural poor with credit is not a goal that is achievable without directly confronting the incentive structures facing lending institutions, borrower groups, and other actors. Sociopolitical considerations are central to these incentive structures. Developing country policymakers frequently subordinate the provision of financial services to political priorities by loan targeting, supporting unprofitable enterprises, or subsidizing interest rates. These factors interact with the technical aspects of rural financial markets to profoundly affect prospects for both institutional and development sustainability. The Indonesia case clearly illustrates these interactions, confirming experiences from other countries.

Over the last fifty years the world's population has more than doubled. By the year 2020 there are likely to be almost 2 billion more people to share the planet. Economic growth has meant that most of us have more money to spend; as a result, global demand for food has expanded hugely. Yet agriculture and the food industry manage to supply ever-larger amounts of food and raw materials to satisfy this demand.

Paradoxically, the enormous increase in production has not solved the problem of food security. Although much progress has been made, roughly 20% of the world's population is still malnourished. Simultaneously, some countries struggle to control production in order to avoid the build-up of costly surpluses of food. The existence of want amidst plenty shows that ensuring global food security — the consistent access to adequate nutrition by all people in the world — is not a simple problem with a simple solution.

In many countries the structural transformation of agriculture over recent decades has been astonishing. Productivity has improved substantially, often at a faster rate than in the economy as a whole. In many parts of the world, agriculture is now a sophisticated industry employing advanced technologies. Many of the labour intensive practices of the past have been replaced by capital-intensive methods, using far more chemicals, water and energy to produce, process and market food. Without the considerable improvements in productivity that these advances have created, it is unlikely that the average OECD consumer would have seen the striking fall in the proportion of disposable income he or she spends on food — now less than 20% in many OECD countries compared to 30-35% at the beginning of the 1960s.

Continued increases in the efficiency with which resources are used in agriculture are essential to promote economic growth and raise living standards, particularly in poorer countries where agriculture is often a major part of the economy. Thus there is nothing to suggest that the forces contributing to the structural transformation of world agriculture will abate. But it is growing clearer that the increasing demand for food is exerting substantial pressure on the environment. The degradation of wildlife habitat reduces biodiversity. The use of large quantities of agro-chemicals damages water quality. Productivity is lost through, soil erosion and salinisation. Any number of similar examples come to mind. In spite of the contribution of agriculture to shaping the landscape and maintaining the countryside, eco-systems are clearly under strain, and the environmental performance of farming will have to be improved.

In order to ensure food security and at the same time the protection of the environment, a number of problems have to be tackled. In so doing, a global perspective is vital. Although food security is a national obligation, it is not one which can be pursued effectively by countries in isolation. International co-operation and coordination will be necessary in the search for solutions.

First and foremost, the world community has to tackle the problem of poverty which lies at the root of much food insecurity. Sustainable economic growth is a crucial element in raising incomes and thus guaranteeing access to food. Ensuring a macroeconomic climate which is conducive to private investment and promoting freer trade are both basic requirements for improving economic well-being. Microeconomic policies should focus on eliminating structural impediments to growth, such as poor transport and communication infrastructure or the lack of skills in the use of new technology. Overcoming these barriers will contribute to job-creation, the development of human capital and a more efficient allocation of resources within and among countries. The higher, sustained rates of growth necessary in developing countries can be met through such policy approaches, complemented by well-targeted development assistance.

Second, the policies which form the economic framework must provide the right incentives to farmers and others for the use of agricultural practices that are economically and environmentally sustainable. In many poorer countries agricultural policies have often taxed producers, thereby depressing production and limiting the adoption of sustainable agricultural practices. In these countries, continued policy reform will be required to redress the balance. Similarly, in many richer countries, the reform of those agricultural policies that promote the intensification of productlon through artificially high prices or subsidies will play a central role in reducing the damage that agriculture does to the environment.

Where there are still environmental problems, farmers can be rewarded directly for reducing harmful effects, protecting the countryside and improving wildlife habitats. Research and development, and the diffusion and adoption of improved techniques, will be crucial in stepping up production to meet demand, while nonetheless safeguarding the environment. Above all, market-oriented agricultural sectors should respond to changes in global supply and demand rather than being isolated from international developments. Indeed, countries should make full use of international markets in ensuring food security. Trade encourages the efficient transfer of food supplies from areas where there are surpluses to others where there are deficits. It allows countries to become self-reliant, instead of wasting large amounts of scarce resources trying to become self-sufficient. In the special circumstances created by food crises in poor countries, the selective and judicious use of food aid is required. But it should be used to make up temporary shortfalls in production, rather than forming a permanent source of domestic supply. It should be of local origin, where possible, rather than being used by richer countries to dispose of surpluses.

The traditional aim of analysis at the OECD has been to identify the most effective and efficient mix of policies, both macro and micro, to achieve both economic and social objectives. Food security is a case in point. Coherent policies, good governance and efficient markets are vital for ensuring national and global food security. Since food insecurity is the product of a complex mixture of physical, social and economic factors, attacking it requires an interdisciplinary approach. The work of the OECD is taking such a perspective in the search for appropriate policies. Achieving food security is a global challenge that requires countries to work together. The work of the OECD, its promotion of international co-operation and its growing emphasis on expanding the policy dialogue with non-member countries will all contribute to finding solutions to the most basic of human needs — the need for food.

ENSURING GLOBAL FOOD SECURITY

Eradicating malnutrition across the globe will require an ambitious and coherent mix of policies, covering agriculture, trade, aid, environmental protection, research, investment and more. The correct balance will help provide the conditions for a general rise in prosperity — which is what puts food in people's mouths.

At the World Food Conference in Rome in 1974, the world's governments examined global food production and consumption, concluding their meeting with the solemn declaration that 'every man, woman and child has the inalienable right to be free from hunger and malnutrition in order to develop their physical and mental faculties'.

In spite of the considerable progress in increasing food production and reducing malnutrition over the two intervening decades, those ambitious words have yet to become a reality: the Food and Agriculture Organisation of the United Nations (FAO) estimates that over 800 million people in the developing world alone are still undernourished. Indeed, the FAO fears that, unless action is taken, many of the current problems of food security will persist and some will become worse. In Africa in particular, lagging food production and a rapidly growing population are increasing the risk of serious and persistent food shortages.

At the World Food Summit held, again in Rome, from 13 to 17 November 1996, the member governments of the FAO reaffirmed their commitment to the eradication of hunger and agreed on a plan of action for increasing the year-round availability of food, improving access to supplies, and reducing the number of malnourished people.

To guarantee that all individuals have access to sufficient food — their 'food security' — the world must produce enough to feed its expanding population; and that output must be readily available to all. It is therefore literally vital that government policies do not impair incentives to produce and market food. Many poorer countries, in attempting to keep costs low for urban consumers, depress the prices received by farmers and actually worsen the supply of food.

Developing Food Resources

With enough investment to support the development and application of improved methods of production, the world can indeed produce the quantities required, even though its population is projected to rise from roughly 5.8 billion currently to 7.6 billion by the year 2020. In recent decades world production of food has expanded at a rapid rate, aided from the 1960s onwards by a scientific revolution that has led to major gains in the yields of rice and wheat.

The introduction of improved varieties of crops and the use of fertiliser, pesticides, mechanisation and irrigation have resulted in production increases in excess of population growth in many regions of the world. Increases in yields per hectare of 3% or more per year have not been uncommon. Science and technology have been of paramount importance, first, by making possible the dietary improvements that have accompanied the expansion of supply, and, more generally, in promoting economic development by lowering the real cost of food and releasing labour from agriculture for employment in other sectors.

In many richer countries, the speed with which agriculture is expected to adjust to changes in productivity has been a contentious political issue. Attempts have been made to protect farm incomes by supporting prices and even to limit necessary structural changes, such as an increase in farm size. Many of these attempts have been costly failures. They may even have accelerated the very pressures for structural change that they were intended to control.

As supply continues to expand, more attention will have to be given to the protection of the environment, which is now coming under considerable pressure from intensive farming techniques. Most of the world's available cropland is currently being used, and little further production can be expected from bringing additional land under cultivation. The increases in crop yields necessary to meet demand will have to be both economically and environmentally sustainable. Further advances in biotechnology, such as the genetic modification of plants to reinforce their resistance to pests or diseases, offer Substantial promise for increasing yields with minimal impact on the environment. Improvements in the management of scarce natural resources, particularly water, are important. Irrigated agriculture contributes nearly 40% of world food production from only 17% of the world's cultivated land, but accounts for 70% of global water use.

Nor are these concerns purely terrestrial. Fish are an important source of protein, particularly in developing countries, with a vital role to play in food security. But stocks are under considerable pressure because of over-fishing. A plan of action for the sustainable exploitation of fisheries was agreed in Kyoto, Japan, in December 1995, by 95 member governments of the FAO.

It calls for improved international co-operation and co-ordination in all aspects of fisheries management, not least better monitoring of stocks, strengthening the scientific basis for their management, and the use of fishery conservation and management arrangements to reduce over-fishing. If such a route is not followed, there is a risk of a large deficit in supply early in the next century. That would hit the developing countries particularly hard.

The expansion of supplies through aquaculture could be important in filling the gap. The FAO estimates that world aquaculture production is likely to more than double by the year 2010, and could then account for over 35% of fish consumed. But so far aquaculture has centred largely on high-value species, such as salmon and shrimp, which are unlikely to provide inexpensive protein for developing countries. Their best option to increase the supply of protein is to focus on the promotion of aquaculture as part of sustainable agricultural systems by, for example, the rearing of fish in rice paddies.

The creation of sustainable agricultural systems is a complex process that requires appropriate technology and policies, and attention to education and training. The effective diffusion of improved techniques is crucial in increasing output and minimising post-harvest losses. And since many of the world's food-producers are women, equal access for both men and women to training in these techniques is extremely important.

The returns from agricultural research are high, yet investment in research, particularly in developing countries, has been stagnant in the 1990s. That calls for a higher priority for funding, both nationally and internationally. Closer co-operation between the private and public sector — through, for example, joint ventures and the protection of intellectual property — would help to ensure that sufficient resources are allocated to research in the future. The OECD countries can help here: their extensive R&D infrastructure and scientific expertise can be used to strengthen agricultural research programmes in poorer countries, particularly through co-operative projects and training.

Economic Growth V. Poverty

Food insecurity can be a vicious circle: people who are malnourished' often do not have the resources to grow their own food, or enough income to purchase it from others. The solution clearly lies in increasing the incomes of the poorest. In some countries that means ensuring that poor people have access to sufficient firmland and capital. In others it means finding ways to promote the growth of economic activity in rural and urban areas. The creation of a policy environment that encourages private investment and capital-formation throughout the economy is an important ingredient in accelerating economic growth and the creation of employment. All governments have to direct their efforts towards ensuring the necessary conditions for sustainable economic growth, which brings with it an improved distribution of income.

Political instability often contributes to food insecurity. Participatory and pluralistic political systems, with governments responsive and responsible to their peoples, are thus most conducive to food security. The disruption of food supplies through wars and civil strife has been a major factor in causing malnutrition and famine.

By contrast, economic growth and political stability in many parts of the world have done much to improve food security. It is not coincidental that where new techniques have made the largest contribution to agricultural productivity, in many of the large rice-producing countries of Asia, is also where the fastest economic growth has been occurring. Other regions of the world, particularly Africa, which have not had such a favourable combination of suitable technologies and political stability, have seen their food supplies deteriorate under the pressure of rapid population growth. It is towards such areas in particular that the focus of improving food security must be directed.

Reforming Agricultural Policies

The world's farmers can feed its growing population if agricultural policies do not distort their incentives. Prices must reflect changes in supply and demand in order to guide the use of resources and stimulate productivity. Attempts by governments to try to insulate domestic markets from changes in world demand and supply are often costly and merely transfer the burden of adjustment to others.

Further reform of agricultural policies and the reduction of government intervention is essential to improve the functioning of markets at all stages — from the supply of inputs to the processing of raw commodities and the distribution of final products to consumers, and from local and regional markets to international ones. The reform of policies in many developing countries which tax production and those in many rich countries which subsidise production is necessary for markets to function efficiently. Lessons drawn from the expensive mistakes of agricultural policies in many OECD countries would help ensure that they are not repeated elsewhere as incomes increase. Concern is sometimes expressed that the opening-up of domestic markets will expose consumers to unacceptable variability in prices and its associated risks. This risk can be handled through private-sector mechanisms, such as futures and options markets, provided that they are allowed to operate efficiently and that producers and consumers alike have access to them. The spread of information' technologies offers considerable promise here. State-subsidised storage of agricultural commodities or other measures, such as restrictions on international trade, are likely to be costly and of limited effectiveness. That is not to say that governments have no role at all: the efficiency with which markets can handle fluctuations in supply and demand could be enhanced through assistance to infrastructural development, such as the improvement of transport and communication systems. It is increasingly recognised that the elimination of impediments to the free flow of goods and information is a key factor in increasing the efficiency with which food markets function.

International trade has a major role to play not only for its physical capacity for moving food from surplus to deficit areas, but also through its contribution to growth in world incomes. The aim of trade policy should be to increase self-reliance through the use of commercial imports and exports to balance changes in domestic production, rather than to pursue self-sufficiency regardless of cost. Trade will be vital in meeting the long-term food requirements of a large portion of the world's population, and in alleviating the short-term pressures of domestic crises. Food aid will obviously be important in such emergencies, but it should take care not to skew the price signals that allow markets to work efficiently.

To ensure that international trade plays its role in food security, the reform of policies that insulate domestic markets from international ones is a high priority. The Uruguay Agreement on agriculture establishes a framework to this end. Members of the World Trade Organisation (WTO), recognising that the achievement of the long-term objective of a substantial and progressive reduction in support and protection would require an ongoing process, have agreed to initiate further negotiations on agriculture, starting in 1999. The aim is global integration of domestic markets.

FAO Special Programme for Food Security

In 1994, the FAO Director-General initiated a review of the Organizations priorities, programmes and strategies, which concluded that improving food security should be reaffirmed as its top priority, and the urgent need for its programmes to focus more sharply on increasing food production, improving stability of supplies and generating rural employment, thus contributing to more accessible supplies. The Director-General also proposed that FAO should launch a Special Programme for Food Security (SPFS), focused on the 86 Low-Income Food-Deficit Countries (LIFDCs)—countries least able to meet their food needs with imports. This approach was endorsed by the 1996 World Food Summit, with a Plan of Action in the following areas: ensuring enabling conditions; improving access to food; producing food; increasing the role of trade; dealing adequately with disaster; and investing in food security.

SPFS commenced its operations in late 1994. A budget of $10 million was provided for the 1996/1997, 1998/1999, and 2000/2001 bienniums. Its main objective is to help LIFDCs to improve food security at the national and household levels, through rapid increases in food production and productivity, by reducing year-to-year variability in production on an economically and environmentally sustainable basis and by improving people's access to food.

The underlying assumption is that in most LIFDCs viable and sustainable means of increasing food availability exist but are not realised because of a range of constraints that prevent farmers from responding to needs and opportunities.

By working with farmers and other stakeholders to identify and resolve such constraints, whether they are of a technical, economic, social, institutional or policy nature, and to demonstrate ways of increasing production, SPFS should open the way for improved productivity and broader food access. Its formulation is a national responsibility, supported as necessary by FAO, and usually carried out by a team of government staff or national consultants, guided by a Steering Committee, bringing together representatives of the public and private sectors, farmers' organizations and in some countries NGOs active in the rural sector.

The South-South Cooperation Scheme under SPFS strengthens cooperation among developing countries at different stages of development, with the support of interested donor countries and FAO. It allows countries to benefit from the experience and expertise of more advanced developing countries, who provide a considerable number of experts for two to three years to work in the recipient countries, directly with farmers in rural communities involved in SPFS. FAO launched the South-South Cooperation Scheme at the beginning of 1997. In addition, arrangements to support SPFS have been made under the programme for technical cooperation between developing countries (TCDC). SPFS is also a vehicle for collaboration between FAO and its development partners. When FAO receives a request from a member country for the initiation of the Special Programme, the organization initiates discussions with interested development partners—bilateral and multilateral—as well as with NGOs and the private sector, through its representation offices at the country level. Where concrete agreements are reached, joint missions are organized at the formulation stage, as well as during the implementation process, to monitor progress and ensure the achievement of the Programmes objectives.

SPFS: Structure and Dimensions: In June 2001, SPFS was operational in 64 countries covering more than two thirds of the countries listed as LIFDC. The regional division is as follows: Africa, 38 (LIFDC coverage, 86%); Asia, 14

(58%); Europe, 2 (67%); Latin America, 8 (86%); and Oceania, 2 (33%). SPFS is in various stages of formulation in another 17 countries. Besides LIFDCs, more and more developing countries are coming forward to implement SPFS and managing to find resources to do so—examples are Zimbabwe, Uganda, Peru and South Africa.

One of the distinguishing features of SPFS is that it aims to link practical demonstration work conducted by a farmers with changes in the institutional and policy environment 5 that address constraints to the wider replication of successful innovations.

For this reason, institutional arrangements have to be made at the local level to ensure a high degree of farmer participation, and at the national level to bring people who are influential in the processes of policy and institutional change together. There is also a clear need for strong and competent technical leadership. How these requirements can best be met will depend on country circumstances.

In some cases, where there is already a large degree of decentralization, mechanisms may already exist that provide for adequate farmer participation in the design and management of local-level programmes. Some countries also have high-level National Food Security Commissions, chaired by the President or Prime Minister. Others are in the process of setting up such bodies as part of their commitment to the World Food Summit's Plan of Action. In such cases, options for building SPFS management into existing structures warrant careful examination. Elsewhere, it may be advantageous to create an entirely new institutional framework for SPFS management and oversight.

This could consist of: An inter-ministerial policy committee: Chaired by a Head of Government, with the participation of the ministers for agriculture, livestock, water resources, forestry, fisheries, planning and finance and other ministers involved in food security matters. It would provide policy guidance, ensure necessary intra- and inter-sectoral coordination between institutions, programmes and projects operating at national level, including external assistance on food security/rural development/poverty alleviation related matters, monitors the implementation of the SPFS National Action Plan and would act upon the policy-related feedback from the pilot activities.

An inter-ministerial technical committee: Consisting of directors of agriculture, fisheries, livestock, forestry, planning, agricultural research, etc., providing technical leadership and ensuring integration and coordination of SPFS with other national programmes and activities undertaken by donors, NGOs and the private sector.

An interdepartmental regional committee, preferably chaired by the head of the region or district, consisting of representatives of local government institutions, agricultural services, health officials, the private sector, NGOs, women's and farmers organizations, etc., providing guidance and supervision to the formulation, implementation and evaluation of the programme, and ensuring its integration with other relevant regional/district activities.

Local implementation committee, established at every SPFS site, consisting of technicians dealing with the four components of the Programme: water control, crop intensification, diversification into short-cycle animals and constraints analysis, as appropriate. It would ensure implementation of SPFS at site level in accordance with the work plan.

Most participating countries have appointed a national project coordinator for SPFS responsible for both formulation and implementation of the Special Programme.

Farmers in Baimamiao village in the Sichuan Province of China have increased the yields of their four major crops by as much as nearly 70 per cent since 1992, thanks to work carried out under SPFS. Sweet potato yields showed the sharpest increase, jumping from an average of 2,085 kilogrammes per hectare in 1992-1994 to 3,540 kilogrammes per hectare in 1997.

Yields for maize, wheat and rice also improved, with increases of 43 percent, 24 percent and 16 percent, respectively. In the past, Baimamiao village was a low-income food-deficit area, where poor soils, inadequate and ageing irrigation systems, and the lack of modern agricultural inputs and practices kept crop production and farmers' incomes low.

Local government, with the support of SPFS, decided to focus on water management and soil improvement to help farmers increase their production and incomes. After three years, farmers and SPFS staff can see the results of their work. Arable land from the valley floor to the hilltop has been rehabilitated. Dry-land has been converted into paddy, and existing paddy land has been reshaped and the soil deepened. The infertile thin soils on the hillside have been converted into terracing. Old irrigation facilities have been repaired and new irrigation and drainage channels have been built.

Mountain ponds have been rehabilitated to increase water storage capacities. Trees have been planted on hilltops. The work on water control and irrigation has had a particularly beneficial effect, reducing vulnerability to drought, wastage of water, water-logging of land and soil erosion. In a recent report from SPFS staff in the area, farmers were quoted as saying: "No more worry about drought and water-logging after building irrigation and drainage ditches in field. Because water storage ponds are built in upland area, water can be stored for emergency use. It is really easy to grow crops now!"

But the improvements did not stop there. Better soil conditions and water management were the foundation for improved farming practices, using improved seed varieties and modern crop production techniques, which include raising seedlings and then transplanting them into the fields, using plastic sheet covers, fertilization with a combination of organic manures and chemical fertilizers, and concerted prevention and control of pests and diseases. Annual per capita farmer's income has more than doubled since improvement work began in 1994, jumping from an average of 667 Yuan in the period 1992-1994 to 1,277 Yuan in 1997.

At the heart of the FAO South-South cooperation initiative is exchange of knowledge and experience between developing countries. In this scheme, more advanced developing nations send experts and technicians to work directly with their counterparts and farmers in other developing countries. An example is the South-South cooperation agreement between Egypt and the United Republic of Tanzania. Egypt's longstanding expertise in irrigated agriculture can be particularly useful to Tanzania. Most of the land under cultivation there is rain-fed, leaving farmers at the mercy of weather patterns and conditions. The first group of twelve Egyptian experts and technicians are in Tanzania working side by side with local farmers on irrigation, plant nutrition and soil fertility, marketing, cooperatives, rural credit and finance, seed production, agricultural engineering, and animal and crop production.

The key to the success of SPFS in Tanzania has been the active participation by stakeholders at every stage. Using participatory techniques such as Participatory Rural Appraisal, various stakeholders—farmers groups, input suppliers, credit institutions, research and extension officers, policy makers and authorities at the regional and district levels—were consulted first to identify and then, at a later stage, estimate aspects of Phase I applications, such as increased production, expertise and human requirements, and collection of information.

Community involvement was such that 78 participatory farmers' groups were formed in Dodoma and Morogoro, pooling the resources of 1,116 men and women in common bank accounts and saving and credit associations (SACAs). Backed by these savings accounts, farmers and women in particular improve their negotiating position towards input suppliers and traders in transport, marketing and temporary credit for inputs, in procuring timely tractor hiring services and marketing p roduce. Rice production has increased in areas with SACAs, since the farmers have had greater access to inputs.

In Kilombero district, for instance, inputs bought by a group of women farmers raised production from 17 bags of paddy per acre to as much as 35 to 40 bags. The Government plans to extend SPFS to all areas of the country as a national special programme. This poses a challenge, and the arrival of experts from Egypt will provide valuable support.

A number of bilateral and multilateral donors are supporting pilot phase activities in countries where SPFS is already under way, as well as the formulation of the programme in newly participating countries. The donors include Belgium, France, Italy, Japan, the Netherlands, Spain, Libya and the United Nations Development Programme. Several financial institutions have shown an interest in collaborating in the implementation of SPFS. The FAO Director-General signed Memoranda of Understanding (MOUs) with the World Bank on 13 January 1997 and with the African Development Bank on 20 January 1997 for the promotion of rural development and food security. Both banks agreed to consider providing loan funding of up to $1.5 million for the pilot activities of SPFS in Africa, subject to requests from the Governments concerned.

Working with the Government of Burkina Faso, a joint FAO/World Bank mission has identified a project for the promotion of private irrigation development, which builds heavily on SPFS experience in demonstrating improved irrigated crop production systems. Negotiations are under way in a number of countries to implement these agreements. Collaboration is also expected with the Islamic Development Bank, with whom the Director-General signed an MOU on 21 June 1997. The Bank has shown interest in supporting SPFS in its least developed member countries.

POTENTIAL BENEFITS IN FOREST MANAGEMENT

Forests are central to human landscapes, and civilizations have vanished because they failed to protect forests that in turn protected fragile soils and water sources. Present-day forest managers have a responsibility to spare California a similar fate. Forests are popularly believed to reduce flooding and increase dry-season base flows, as romanticized in the novel "The Man Who Planted Trees". The scientific evidence for relations between forests and water supply, however, has not been as compelling. Research has shown that forests have had a limited role in flood protection and variable effects on total water yields and base flows. In contrast, several studies have convincingly demonstrated that forests protect water quality by protecting the land surface from erosion and filtering pollutants.

Winter snowpacks presently store enormous quantities of water in the forested watersheds of California. Water stored as snow is released slowly during spring snowmelt, in constrast to rainfall that runs rapidly to streams. Snowpacks therefore supply water when it is most needed by humans and the environment, and reduce the need for additional downstream dams and reservoirs.

The predicted shift toward more rain and less snow is of critical importance for water management because the existing water-development infrastructure relies on regulation of streamflow through gradual release of water during snowmelt. As snow is replaced by rain at mid-elevations, winter flood peaks will become larger and more frequent, and reservoir storage is likely to be exceeded in wet months when demand is low. Summer baseflows will be correspondingly lower when demand is high. These climate-driven impacts will likely lead to proposals for many new dams and reservoirs on forest streams, with their resulting environmental impacts. Climate changes directly affect forests through increased drought stress, which makes trees more vulnerable to insect attack. Wildfires are also likely to increase in frequency, size, and severity as climate warms. These stresses on forests will affect their capacity to naturally regulate streamflow and buffer water quality. Many streams that are now perennial are likely to become intermittent with the resulting loss of riparian zones, aquatic habitats, and other beneficial uses of water that depend on perennial flows. Recent concerns with increased demand for water, extended drought, economic and environmental costs of new water-supply infrastructure, effects of water transfers on endangered species, and effects of climate change on water supply have raised the importance of forest management for protection and improvement of water resources. New research and public perceptions may provide opportunities that were not available in the past to forest and water managers. Although the current scientific consensus supports the perception of forests primarily as protectors of water quality, the potential for improvements in the availability of water should not be overlooked. Forest management activities that alter streamflow regimen to benefit downstream water users may be more successful than attempts to increase total water yield. The following sections discuss forest management actions that have potential for improving water resources in California.

MEADOW GROUNDWATER STORAGE

Meadows within the forests of California are small alluvial landforms that are located along streams in the mountainous headwaters of the state. Owing to geologic factors that favor meadow formation, most of California's meadows are in the Sierra Nevada, which has more than 10,000 meadows comprising a total of roughly 300,000 acres. In contrast to the surrounding terrain, meadows have gentle slopes and support flood-tolerant herbaceous plants and shrubs rather than conifers. Under natural conditions, meadows lack incised stream channels, and high flows spread across the meadow surfaces and recharge meadow aquifers. Meadows with intact vegetative cover act as natural reservoirs, regulating streamflow regimen through storage and release of snowmelt and rainfall runoff that passes over and through fine-grained, sod-covered meadow deposits. These meadows attenuate flood peaks and prolong dry-season base flows. The importance of meadows in regulating streamflow is likely to increase as climate change results in a shift from snowmelt to rainfall-dominated runoff at mid-elevations in the Sierra Nevada.

Meadows also provide critical aquatic and riparian habitats for fish and wildlife. Montane meadows of the Sierra Nevada are very important to amphibians such as the mountain yellow-legged frog and birds including the willow flycatcher and Bell's vireo that utilize nesting habitats in meadows to successfully rear young. Kattelmann and Embry (1996) stated that about 20%

of the 400 Sierra Nevada terrestrial vertebrates are dependent on riparian areas. Field studies have shown that the health of benthic macroinvertebrate and fish communities is strongly correlated with riparian conditions. Most of the meadows throughout the Sierra Nevada were eroded by incised channels, or gullies, prior to 1940 as a result of early unrestricted livestock grazing, roadbuilding, railroad construction, elimination of beavers, and other causes. Although current activities on national forests are carefully managed to avoid damage to meadows, the effects of earlier practices remain on the landscape and will not heal without active restoration.

Eroded meadows lose their capacity to store and release water. Gullies convey and concentrate flood peaks more rapidly than well-vegetated meadow surfaces, and therefore aggravate downstream flooding and reduce recharge of meadow aquifers (Liang and others, 2006; Hammersmark and others, in review). Gullies drain groundwater stored in meadows, decreasing the amount of water available to sustain streams during dry summer months, although reduced evapotranspiration may offset drainage to an unknown extent. Channel erosion in meadows adds to stream sediment loads through bank erosion and headcut retreat, adversely affecting downstream water quality and reservoir capacity.

Erosion of meadows adversely affects aquatic and riparian habitats by altering meadow hydrology. Fish passage is not possible through incised channels that are dry for much of the summer months, and the reduced discharge of cold water stored in meadows is likely to increase downstream water temperatures, to the detriment of cold-water aquatic species such as trout.

Wildlife species that depend on wet-meadow plant communities are displaced when meadow water tables drop below rooting depths. According to the Sierra Nevada Ecosystem Project (1996), aquatic and riparian habitats are the most altered and impaired systems in the Sierra Nevada Province.

Drying of meadow soils allows invasion by drought-tolerant brush and conifer species that contribute to heavy fuels loading and add to the risk of catastrophic wildfires.

Loss of high-quality forage provided by wet-meadow sedges, rushes, and grasses decreases forage value for meadows that are grazed by livestock. Meadow restoration is a form of groundwater banking that can provide a wide array of ecological benefits in addition to enhancing water supplies. California's forests encompass the headwaters of the major rivers within the Sierra Nevada, and include thousands of meadows. A regional approach to meadow restoration could help to meet the State's needs for high-quality water and aquatic habitat and help offset the effects of climate change.

The USFS manages many Sierran meadows on national forest system lands, and has been actively working with partner agencies and organizations to restore the hydrologic, geomorphic, and biologic functions of meadows damaged by channel downcutting since 1940. Several projects using the new "plug and pond" approach have been successfully implemented in the past 10 years in the Shasta-Trinity, Plumas, Tahoe, and Sequioa National Forests, as well as the Lake Tahoe Basin Management Unit.

The "plug and pond" method involves preventing headcut migration, filling or plugging gullies, routing surface flows over the meadow surface, and raising meadow water tables. In most cases, livestock grazing has been temporarily halted during and after restoration projects. Permanent exclusion of livestock is not generally necessary to protect riparian resources if pastures are effectively managed to limit cattle numbers, distribution, and seasons of use. Like dams, meadow restoration does not create "new" water, but alters the temporal distribution of streamflow so that less water flows downstream during peak runoff periods in the winter and spring, when water is not in demand, and more more is released during the summer low-flow season, when demand is high. Based on the limited available information and a reasonable range of assumptions, meadow restoration in the Sierra Nevada could increase the amount of groundwater stored in meadows by 50,000 to 500,000 ac-ft annually. The wide range in these estimates results from uncertainties in channel depths and specific yields of meadow alluvium. Increased groundwater storage in meadows would be likely to enhance summertime instream flows, a function that will become increasingly important owing to climate change.

Meadows are often considered to be "sponges" that absorb and release water. This analogy is appropriate for meadows that are supplied by streams and local snowmelt. Meadows that are supplied by groundwater flowing through surrounding hillslopes and underlying fractured bedrock, however, may more accurately be characterized as "valves" that regulate the discharge of regional groundwater as it flows through relatively low-permeability, organic-rich, fine-grained alluvium. Meadow restoration is likely to be most effective in prolonging the duration of base flows in meadows that act as "valves" in regional groundwater flow systems. On the basis of bedrock permeability , meadows that function as "valves" are more likely to be found in volcanic and weathered granitic watersheds than in glaciated granitic watersheds.

Meadow restoration offers a number of other benefits in addition to improving summertime streamflows . Some of these benefits are directly related to water resources, including attenuation of downstream flooding, improvements in downstream water quality, and reduction of reservoir siltation.

Other benefits are related to natural resources other than water, and include improvements in fish and wildlife habitat, protection of sensitive species, reduction of fuels loading, carbon sequestration, and improvements in range conditions. Alluvial valleys in mountainous areas throughout the world, including Africa, Australia, Europe, and South America, are faced with erosion and water-supply problems similar to those facing California's Sierra Nevada montane meadows. Many of these alluvial valleys provide water, crops, and forage that sustain local communities and economies. Successful restoration of meadows in California could provide methodologies that are applicable to critical land and water degradation problems around the world.

Riparian Forests

Riparian forests are forested lands that are located immediately adjacent to streams, lakes, or other waterbodies. Riparian areas represent a transitional zone between aquatic and terrestrial habitats. These habitats are related to and influenced by surface or subsurface waters, especially the margins of streams, lakes, ponds, wetlands, and seeps. Boundaries between riparian and upland forests are not always distinct. The spatial extent of riparian areas varies laterally throughout the channel network and is strongly influenced by geomorphology. Riparian floodplains in forested watersheds may affect or be affected by hydrologic processes in several ways. Due to flow resistance associated with a forested stand condition, flood flow velocities are reduced considerably.

Additionally, when flow on the floodplain occurs, there is a large increase in the cross-sectional flow area. Once a stream rises above the floodplain surface, the slope of the line defining the relationship between river stage and discharge markedly decreases due to increased surface roughness and shallow water depth. In most instances, both the velocity and depth of water flowing outside the channel declines with distance away from the channel. This allows flood waters to recharge alluvial groundwater aquifers and attenuates downstream flood flows.

Floodplains are very important in providing habitat for riparian-dependent species. Floodplains are zones of very high biological diversity, having the highest biodiversity for both terrestrial and aquatic species of any part of the landscape at the watershed scale. In light of this high biodiversity, the California Forest Practice Rules require that native aquatic and riparian associated species and the beneficial functions of riparian zones must be maintained where they are in good condition, protected where they are threatened, and restored where they are impaired (in so far as is feasible). Harvest of coast redwoods on riparian floodplains in the northern Coast Range has been controversial because redwoods on alluvial flats have both high timber volume and unique ecological benefits.

Studies have shown that riparian forests can improve water quality. As surface runoff passes through riparian forests, materials carried in suspension are filtered by riparian trees that act as buffers to reduce the amount of sediment, nutrients, and pesticides entering water courses. Riparian trees also lower stream water temperatures through canopy shading. Shade is important for many fish species that can be adversely affected by elevated water temperatures at different life stages. In addition, riparian forests deliver wood to streams that contribute to pool formation and provide cover for fish. Riparian forests are protected on federal, state, and private timberlands through riparian buffers that limit management actions such as timber harvesting and roadbuilding near streams. The width of riparian buffers, and restrictions on management activities within them, are based largely on land ownership. Within the national forests, riparian buffer widths vary based on provincial plan standards and guidelines. Riparian protection is most extensive for the 6 forests that operate under the Northwest Forest Plan. Fuels reduction within riparian buffers may be needed in some cases to reduce threats of catastrophic wildfires. Removal of trees from riparian buffers remains highly controversial, and forest management and regulatory agencies are carefully evaluating monitoring data particularly with regards to the use of mechanical equipment in streamside zones.

Some riparian forests are used for livestock grazing, usually within allotments that are primarily upland pasture. The availability of water and forage make riparian areas attractive to livestock, which can damage riparian forests through trampling, browsing, and contamination of streams with fecal material.

BMPs for range management and national forest standards and guidelines for riparian management are designed to protect riparian forests from damage by livestock. Although exclusion of cattle may be needed during and immediately after restoration of riparian forests, grazing strategies that minimize impacts on riparian forests through livestock numbers, distribution, and season of use can be used to eliminate the need for permanent fencing. Riparian forests include phreatophytes (trees dependent on availability of shallow groundwater, such as aspens, cottonwoods, and willows) as well as upland species such as oaks, maples, and conifers that utilize moisture above the water table. Riparian forests therefore may affect and be affected by channel incision and groundwater storage in much the same ways as meadows. Riparian trees can also be directly impacted by changes in channel geomorphology, including bed aggradation and channel widening that are frequently linked to upstream land uses, including forest harvesting and roadbuilding.

The water quantity and quality benefits provided by riparian forests can be preserved and enhanced through actions that maintain natural channel geomorphology. Protection of riparian forests therefore depends heavily on effective management of upland watersheds to prevent excessive runoff and sedimentation.

Vegetation Management

Management of forest vegetation to improve water supplies has a long history in the western United States. Early efforts attempted to reduce transpiration or increase snowpack by removal of trees. Most of these efforts had limited success. Previous studies have shown that increases in water yields resulting from vegetation management are highly variable and difficult to measure, and that treatments must remove at least 20% of the vegetation to have a measurable effect on streamflow. Using computer simulation work, Troendle and others (2007) reported that an increase of 1 acre-foot of runoff theoretically could result for every 12 acres of forest thinning (fuels reduction). They suggested that the water yield response to large scale forest thinning in the northern Sierra Nevada forests would be short lived with a single treatment (perhaps only 15 years), but that an active management programme perpetuated over time could result in subtle increases in water yield. Studies have provided limited evidence that measurable water-yield increases have occurred in larger watersheds in the past. For example, Blanchard (1962—as cited by Zinke, 1987) calculated the cumulative effect of 30 years of logging on the South and Middle Forks of the Mokelumne River in the central Sierra Nevada. He reported that from 1930 to 1961, water yields from these watersheds gradually increased. Approximately 40,000 acres were logged during that time period.

Innovative approaches that utilize selective thinning of younger, smaller trees show some promise for limited improvement in streamflow regimen, as well as reducing fuel loading and increasing carbon sequestration. These treatments, however, also have potential to increase surface runoff and erosion from disturbed soils.

Fuels/Fire Management

Wildfires affect water resources through removal of vegetation and alteration of soils and ground cover. Removal of forest canopies by fire temporarily reduces transpiration and interception losses, and may increase streamflow until vegetative regrowth increases transpiration to or above pre-fire rates.

Burning of schlerophyllous vegetation can lead to development of hydrophobic soils that have reduced infiltration rates. Hydrophobic soils, in conjunction with the lack of ground cover remaining after fires, can increase surface runoff and erosion and reduce infiltration and baseflows. The effects of wildfires on infiltration and runoff are positively related to fire severity. Wildfires can also temporarily affect municipal water supplies when large quantities of water are needed to fight fires in wildland-urban interface areas. Although increased water yield is a potential impact of large, intense wildfires, it is generally not significant. Where 75 percent to 100 percent of the vegetative cover is removed, runoff may increase from 0.1 acre-foot per acre of burned watershed for basins receiving 15 inches of mean annual precipitation to 0.8 acre-foot per acre burned for watersheds receiving 40 inches of mean annual precipitation). In forested areas, water-yield increases are minimal until basal area loss to fire exceeds 50 percent.

The additional water yields that result from catastrophic wildfires are generally considered to have little value for water supply and hydroelectric energy generation. Almost all of the additional runoff occurs during the wet season and must be regulated for dry season use by surface reservoir storage. Typically, flows increase during large storm events when water is intentionally allowed to pass through reservoirs owing to flood-management concerns. Additionally, the additional water yield does not contribute to a dependable water supply, since the additional water is only a very temporary increase in yield. The occasional, short-term positive gains from increased water yield are more than offset by the frequent short and long-term negative impacts of increased peak flows, increased sedimentation and decreased water quality.

High fuel loads that develop in the absence of fire or fuel-reduction treatments eventually lead to catastrophic high-intensity stand-replacing fires that generate large volumes of eroded soil and ash, as well as nutrients such as nitrate nitrogen, ammonium nitrogen and phosphate phosphorus. Fire exclusion can also lead to negative effects on water quality owing to unnaturally large accumulations of forest litter that increase concentrations of nutrients in runoff from forests.

Fuels reduction projects, including thinning, mastication, and prescribed fires, are used to decrease the intensity, extent, and negative consequences of wildland fire. Forest management to reduce fire severity on national forests is currently administered under the National Fire Plan (NFP) and the Healthy Forest Initiative (HFI). About 70% of the 20,000,000 acres of national forest system lands in California, or 14,000,000 acres, are in need of treatments to reduce fuel loads to natural levels. The USFS is currently treating about 200,000 acres per year, while on average 400,000 acres are lost annually to wildfires (Rob Griffith, USFS, personal commun., 2008). The most effective fuels reduction treatments for reducing the spread and intensity of wildfires have been combinations of mechanical treatments and prescribed burning.

Fuel reduction projects can have adverse effects on water quality, but these effects are generally minor and temporary, and are far exceeded by the adverse effects of catastrophic wildfires. The adverse impacts of wildfire are generally much greater per unit of affected area than the impacts of fuels reduction projects, and also affect much larger areas than those included in fuels reduction treatments. Firefighting tactics are increasingly being modified to protect water quality and aquatic organisms. Rapid restoration of areas disturbed by fire suppression actions is routinely included in suppression efforts. Following fire containment, burned areas are assessed and, if necessary, treated to prevent adverse effects on water quality and other resources. The USFS uses its Burned Area Emergency Response (BAER) programme to restore areas burned by wildfires, while CAL FIRE participates in the State Emergency Assessment Team (SEAT) programme on state and private lands. These programmes focus primarily on erosion prevention, as well as the re-establishment of forest vegetation. The BAER and SEAT programmes will become increasingly important owing to projections of higher frequency and intensity of wildfires related to climate change.

Road Management

Watershed research and monitoring work in California have shown that unpaved roads and road stream crossings are usually the dominant source of management-related sedimentation in forested environments. Past work has shown that a relatively small proportion of the total road length produces most of the road-related sediment delivered to streams. Road-related sediment-transport processes include surface erosion, gullying, and mass wasting. Sediment may be delivered to streams either episodically when roads or road-stream crossings catastrophically fail, or chronically due to incremental surface erosion. Excessive sedimentation associated with roads is of concern because of potential impacts on stream habitat and water quality. Thousands of miles of roads have been constructed through forests in California, primarily to provide access for timber harvests.

The eighteen National Forests in California alone contain about 50,000 miles of forest roads. Of these, roughly 20,000 miles may no longer be needed for their original purposes. Private forestlands contain many additional thousands of miles. Forest roads intercept surface and subsurface flows, adding to flood peaks and decreasing recession flows. Stream crossings are vulnerable to damage by high flows and can divert streams from their natural courses, resulting in serious erosion and water-quality problems.

Many of the roads on private timberlands were built decades ago to very low design standards, often in environmentally sensitive locations such as unstable hillslopes and riparian areas. A significant number of the older roads are part of the current road network, while others were neglected and abandoned with no consideration of ongoing erosional impacts . These "legacy" roads are particularly susceptible to catastrophic failure during high magnitude, low frequency storm events, such as the one in 1997 that caused extensive flooding throughout a large part of northern and central California (Furniss and others, 1998, Madej 2001).

Similarly, much of the existing national forest road network was constructed under standards less protective of water resources than present standards, and maintenance of roads has not been completely effective at preventing runoff interception and erosion. Many roads constructed for timber harvest are no longer needed for their original purpose. Adverse hydrologic and water-quality impacts of roads can be reduced by upgrading and replacing culverts, outsloping road treads, and installation of waterbars and grade dips at stream crossings. Roads no longer necessary for resource management or recreation can be effectively decommissioned by removal of fills at stream crossings and partial or total outsloping of road treads, including cuts and fills.

Detailed field surveys are the main tool available to identify the road segments of greatest concern. Public and private landowners in California are actively inventorying their road networks, prioritizing road segments requiring road improvement or decommissioning work, and completing projects. A considerable amount of road upgrade work has been completed to date with both public and private financing. While there are short-term impacts associated with road decommissioning, particularly at road-stream crossings, road treatments will reduce the long-term sediment production from older roads. Improved operator practices are required for road-stream crossing installation at high-risk sites and at decommissioned crossing sites.

Urban Forestry

Trees planted in city parks, lots, private residences, and along streets collectively form urban forests. Although urban forests are not managed specifically for natural resource production or conservation, they have environmental benefits that extend well beyond aesthetics. Trees reduce stormwater runoff by intercepting rainwater on leaves, branches, and tree trunks. Rainwater intercepted by trees soaks into the ground, reducing peak flows, surface runoff, erosion, and contaminant transport.

Canopy interception from urban trees changes runoff quantity and pollutant loads in several ways:

1) Evapotranspiration – increases soil water storage potential;
2) Interception – reduces the volume and timing of runoff;
3) Infiltration – The root system of trees can increase soil infiltration rates;
4) Erosion – Through interception trees reduce the soil impacts from raindrops.

Trees in combination with other types of landscaping can be very effective in intercepting rainfall and filtering pollutants. Comprehensive urban forest management can lead to improved stormwater control and promote efficient water use in urban areas.

3

Participation Approach of Social Forest Management

PAST ACTIVITIES

Forest extension activities were formally launched in the country in the year of 1964 with the establishment of two forest extension divisions at Dhaka and Rajshahi and later two divisions at Comilla and Jessore. It was really a very small programme and the activities were confined only to establish nursery in the districts headquarter and raised seedling and sell the same to individuals and organizations. The location of this programme was so urbanized and limited that it only partially served the needs of the effluent town dwellers only.

BETAGI-POMORA COMUNITY FORESTRY PROJECT

The first community forestry programme in the country, started at Betagi and Pomora mouza (village) under the district of Chittagong in the year of 1979 with the personnal initiative of Prof. A. Alim, renowned forester and Prof. Dr. Mohammed Yunus, founder of Gramen Bank. Initially the project covered 160 ha of Government denuded hilly land at Betagi and with 83 landless participants from adjacent community and subsequently extend over another 205 ha of Government owned denuded hilly land at Pomora with another batch of 243 landless (families) participants. Under this programme each landless participant was provided with 1.62 ha of land for growing tree and hoticultural crops with technical and financial assistance from the Forest Department. This community programme has given the landless an identity of their own and a sense of direction in life. But this model has not been replicated in the other areas due to lack of initiative of the Forest Department as well as the Government.

Rehabilitation of Jhumia Families

Another project was undertaken by the Forest Department in the Hill tract areas to establish plantation through rehabilitation of Jhumia families in 1980. Main objectives of the programme were (i) to rehabilitate tribal families in the

Unclassed State Forest(USF) lands along with rehabilitation of denuded USP land; (ii) to introduce a sustainable agroforestry production system; (iii) to improve the socio-economic condition of the tribal people and (iv) to motivate tribal people in development of forestry.

Under this programme each family was allocated 2.02 ha of USF land for growing agricultural crops (over 1.20 ha), raising plantation (0.80 ha) and for house construction (0.20 ha). The rehabilitated families were given land use rights and were allowed to enjoy 100% benefits accrued to those lands.

The participants were given input support for growing agriculture, horticulture and forestry crops and cash support for house construction. This programme continues for quite a long period of time but could not sustain mainly because of nomadic character of the tribal groups.

Another reason of failure was that the families were rehabilitated in clustered villages without considering their cultural and religious values. Thus in most of the cases it was found that the families have left the area.

A parallel programme was also initiated by the Chittagong Hill Tract Development Board in which Forest Department was responsible for implementation of afforestation component where Cittagong Hill Tract Development Board was responsible for the rehabilitation component. This programme was also not found so much responsive to hilly people except for some plantation establishment.

Development of Community Forests Project

The activities of the first phase of this project began in 1981 and were completed in 1987 in seven greater districts of the North-Western zone of the country.

The main components of the project were:

- Strip plantations along roads and highways, railways, canal sides, district and Union Parishad roads, totalling about 4,000 km.
- Fuelwood plantation on 4800 ha of depleted Government land on participatory concept.
- Agroforestry demonstration farms over 120 ha also with participatory concept.
- Replenishment of depleted homestead wood lots in 4,650 villages.
- Training of Forest Department Personnel and Village leaders.

Development of Forest Extension Services (1980-1987)

Development of Forest Extension Services (Phase II) began in 1980 with the Government funding and subsequently amalgamated in some areas (*i.e.* North-North West district) with Asian Development Bank funded Community Forestry Project. The main activities under this programme were:

- afforestation in some 3100 villages.

- roadside tree planting along 3600 km of primary highways and roads and about 600 km of Union Parishad roads.
- production of 49 million seedlings for distribution.

Thana Afforestation and Nursery Development Project

This project is a follow-up of Development of Community Forestry Project and Forest Extension Project and has been designed primarily to:

(i) Increase the production of biomass fuels and

(ii) Enhance the institutional capability of FD and local administration in implementing a self-sustaining nationwide social forestry programme.

In order to increase the production of biomass fuel and to arrest the depletion of tree resources, the project envisaged to develop tree resources base through planting of depleted sal forest as well as brining all suitable and available land in the rural areas under tree cover with active participation of the rural poor of the locality.

Originally the project was to be implemented by the Forest Department and former Thana Parishad during the period of 1987 to 1994. But in 1992 Government decided that the all project activities were to be implemented by Forest department alone. The major components of the project were:

- Establishment of plantation over 20,225 ha depleted Sal forest areas.
- Development of agroforestry over 4,200 ha in the Sal forest lands.
- Raising strip plantation on 17,272 km along Road and highway, Railways, Embankment and Feeder Roads.
- Raising 1,282 ha plantation in the land outside the BWDB.
- Planting 7.017 million seedlings at the premises of different education, religious and social institutions
- Establishment of 345 nurseries at Thana headquarters.
- Raising of 10.618 million seedling for distribution to public.
- Training of some 76,000 people of different levels.

Here this may be mentioned that at the last stage of the project implementation, the Government has found that this was quite impossible to protect the strip plantation and also impossible to trained 76,000 people by the Forest Department alone.

The Government invited NGOs to participate in this programme for successful implementation. PROSHIKA, POUSH, GRAMMEN BANK and other NGOs came forward to help the Government for successful completion of the project; NGOs employed their group members to protect the strip plantation and ADAB came forward to train people at different levels with the help of its member organisations.

The above plantation activities were carried out with the direct participation of the local people with the help of the NGOs by executing benefit sharing agreement.

Coastal Greenbelt Project

Another project financed by Asian Development Bank is under implementation in the Coastal region of Bangladesh. The main objective of the project is to create a vegetative belt all along the coast to save the lives and properties of the people living in the coastal areas from devastated cyclone and tidal surges which occur very frequently in those areas. All of the activities of this project are also being carried out following participatory approach. In this project also the participants have been selected among the poor people living in the adjacent areas by involving NGO and a pre-designed benefit sharing agreements also being executed with the participants to protect their rights over plantations and to ensure benefit expected to be received out of the plantation.

Agroforestry Research Project

Pilot Agroforestry Research and Demonstration was implemented by the FD in the Sal forest areas. The project had been developed precisely to design/develop agroforestry modules which is environmentally feasible, socio-economically acceptable enhance tree and crop production at the same time to uplift the socio-economic condition of the participants. The project aimed at using 120 ha of encroached Sal Forest land of Dhaka, Mymenshing and Tangail Forest Division to develop suitable participatory plantation models.

Food Assisted Social Forestry Programme

The World Food Programme assisted the Government to develop Social Forestry as a national programme and the Government incorporated WFP assisted social forestry programme in its annual development plan from 1998. Poverty alleviation, economic rehabilitation of rural poor especially the destitute women of the society by engaging them in forestry activities, social uplift of rural poor and environmental improvement are the main objectives of this project. Historically this programme was conceived in the country since 1989 on pilot basis allocating in kind resources (Wheat) to a limited number of NGOs for raising strip plantation along roads, embankments, Highways etc. in rural areas following the participatory mechanism. In implementing this programme FD was involved later on to provide technical guidance to the NGOs and other GOB agencies. At present probably this is the largest Participatory Forestry Programme in Bangladesh. From 1990, 100 NGOs are involved in this programme and at present about 60 NGOs are continuing with the programme. Commencing from 1990 up to 1998 about 31 million trees were planted involving 0.062 million people directly and 0.62 million people indirectly. The programme has created employment to the tune of 68 million man days. This programme is being implemented by the NGOs through contractual benefit sharing among participating poor men & women 60%, NGOs 10%, the rest land owners.

ESTABLISHMENT OF FOREST DEPARTMENTS

The introduction of scientific forest management and the organization of forest departments in India owe their inception to Dr. Dietrich Brandis, a professor of botany from the University of Bonn, who was engaged by the East India Company in 1856 as the superintendent of teak forests in Burma. His resourcefulness, ability and consummate skill led to his appointment in 1864 as the first Inspector General of Forests for India and as a result of his initiative forest departments were constituted in the then British provinces.

A Forest Act, later replaced by the more elaborate Acts of 1878 and 1927, was passed in 1856 and in accordance with its provisions valuable timber-bearing regions were declared as either "reserved," "protected," or "unclassed" forests. In the main the difference lies in the degree of control prescribed. In especially valuable areas, declared as reserved because of the vital role they play in the national economy, only welldefined and limited private rights are recognized.

In protected forests, private rights are admitted freely and restricted only in the interest of the right-holders themselves. In such forests, while the reasonable needs of the local population are met, the local people are not permitted to exercise rights and privileges in blissful disregard of the interests of the generations to come. Stated succinctly, the distinction consists in this: in a reserved forest everything is treated as an offense which is not permitted, whereas in a protected forest nothing is an offense unless it is prohibited. The "unclassed" forests, as their name signifies, are those awaiting classification as either "reserved" or "protected" While the forest departments in the former British provinces of India were highly organized, such was not generally the case in the Indian States, where forests attracted attention more for the sport that they provided than for their protective and productive functions. Most of the states were too small and poor to afford full-fledged forest organizations and the bulk of their forests were far from valuable. The integration brought about by recent constitutional changes since the dawn of independence has now made it possible for each group of states to have its own forest department.

PRIVATE FORESTS

During recent years attention has been focused upon the need for protecting private forests from the short-sighted policies of their owners, as a means of conserving an essential national economic resource. With no prospect of quick annual returns, an owner is frequently tempted to sacrifice his capital for an immediate gain. The provisions of the Indian Forest Act of 1927 did not suffice to arrest the destruction of these forests and the consequent physical deterioration of the areas in which they lie. Recent legislation for the control of private forests, however, has sought to establish the following procedure:

- Designation by the government of private forest areas over which control is to be exercised;

- Issue of felling permits, pending compilation of working plans for the areas so designated;
- Affording the owners of the private forests concerned the opportunity of managing them in accordance with approved working plans;
- Vesting control of private forests in the government in cases where recalcitrant owners indulge in reckless fellings in flagrant disregard of the working plans. While the title to such "vested" forests continues unaffected at present, existing trends point to the complete abolition of the private ownership of forests.

FOREST POLICY

1894. Although the foundations of regular forest management were laid in the sixties of the nineteenth century, it was not until 1894 that the Government of India adopted the forest policy that constitutes the Magna Carta for the forests of India. Since then the vital role of the forests in relation to agriculture, industry, transportation and defence, has received increasing recognition. Their protective functions have also come to be better understood.

The old forest policy was designed largely for the former British provinces. Its provision for the relinquishment of forest land "without hesitation" for the extension of agriculture-subject to certain conditions honored chiefly in the breach-led to undesirable consequences.

In the original policy no mention was made of sustained yields, working plans, forest education, or forest research; nor was any cognizance taken of wildlife, an important and attractive feature of India's forests.

THE NEW POLICY

The recent constitutional changes have directed attention to the shaping of an over-all policy for the country. The government has revised the forest policy in relation to the progress that has been made in the physical, economic and political fields. The New Policy of 1952 provides, among other things, for the classification of forests into:

- Protection Forests, the preservation of which is directed by purely physical and climatic considerations;
- National Forests, to be maintained and managed to meet the needs of defence, industry and transportation;
- Village Forests, to be maintained for their role in the agricultural economy of the country, such as provision of firewood and of small timber for agricultural implements and dwellings;
- Treelands, or areas covered with tree growth of some sort, which, though outside the scope of regular forest management, are essential to the amelioration of the physical environment. These classes are by no means mutually exclusive, since each forest usually fulfills functions subsidiary to those specifically assigned to it.

The notion, once widely entertained, that forests as such have no inherent right to land, but may be permitted on sufferance on residual land not required for other purposes, is being effectively dispelled. In the New Policy, the protective and productive role of the forests finds recognition in a directive to the effect that a proportion of the whole land area to be determined by each state should be permanently maintained under forest. Each group of villages is to have a complement of forest, constituting a fuel, fodder and timber reserve.

Under the New Policy, moreover, attention is given to such matters as:

- Balanced land use; putting each type of land to such use that it will produce most and deteriorate least;
- State control of private forests;
- Preservation of wildlife;
- Control of grazing;
- Control of shifting cultivation;
- Provision against fluctuating budgets.

Special provision is also made for the training of forest personnel, for forest legislation and for research. The principle of sustained yield and the need of setting up an organization for dealing with working plans have both been especially stressed.

PEASANT MOVEMENTS AGAINST STATE FOREST POLICIES

There were major peasant movements against state forest policies in 1904, 1906, 1916, 1921, 1930, and 1942. These recurrent conflicts, remarked one sensitive official, were a consequence of "the struggle for existence between the villagers and the Forest Department; the former to live, the latter to show a surplus and what the department looks on as efficient forest management." The most significant forest movement in Kumaun and Garhwal took place in 1921. This took the shape of labour strikes, which crippled the administration, and the widespread burning of pine forests. A total of 395 recorded fires burnt an estimated 246,000 acres of forest. Hundreds of thousands of resin channels were destroyed.

Constituting a direct challenge to the state to relax its control over forest areas, these protests enjoyed enormous popular support, which made it virtually impossible for the administration to detect the people responsible for the fires. The fires were generally directed at areas where the state was at its most vulnerable, for example, compact blocks of chir forest worked for timber and/ or resin. Significantly, there is no evidence that the large areas of broad-leaved forest, also controlled by the state, were at all affected. Thus arson was not random but carefully discriminating-it spared those species more useful to the village economy.

In the vanguard of the 1921 movement were soldiers who had fought for the British in the First World War. Kumaun and Garhwal had long supplied hardy and exceptionally brave soldiers for the British Army-indeed, three of the five Victoria Crosses awarded to Indians between 1914 and 1918 went to this region.

These former men in uniform saw the forest regulations as a bitter betrayal of their interests by the white overlord for whom they had so recently risked their lives. Their protests alarmed the colonial state, for apart from being a reservoir of able-bodied men whom it hoped to continue to recruit for the wars it had to fight, the Kumaun hills bordered both Nepal and Tibet-regions not under direct British suzerainty but in which it had strong trading and political interests. In the wake of the popular protests, a magisterial critique of government forest policy was published by Govind Ballabh Pant. Pant was a rising lawyer from a peasant household in Almora who went on to become one of the foremost of Indian nationalists, after independence taking office successively as chief minister of Uttar Pradesh and home minister of the Government of India.

His 1922 booklet The Forest Problem in Kumaun described the "burial of the immemorial and indefeasible rights of the people of Kumaun," buried, that is, "between the property-grabbing zeal of the revenue officers and the exhortations of experts of the forest department." As he put it, with legal precision, "the policy of the Forest Department can be summed up in two words, namely, encroachment and exploitation." Several decades of a single-minded commercial forestry had led to a manifest deterioration of the agrarian economy: "Symptoms of decay are unmistakably visible in many a village: buildings are tottering, houses are deserted, population has dwindled and assessed land has gone out of cultivation since the policy of [forest] reservation was initiated.... Cattle have become weakened and emaciated and dairy produce is growing scarce every day: while in former times one could get any amount of milk and other varieties for the mere asking, now occasions are not rare when one cannot obtain it in the villages, for any price for the simple reason that it is not produced there at all."

Pant's analysis was rooted in a deep knowledge of the local context. He took it upon himself to combat the charge, commonly levied against the hill peasant, "of reckless devastation," a charge "sedulously propagated by prejudiced or ignorant persons." As he wrote, The spacious wooded areas extending over the mountain ranges and hill sides bear testimony to the care bestowed by the successive generations of the Kumaonies. All of them are not of spontaneous growth and specially the finer varieties bespeak his labour and instinct for the plantation and preservation of the forest. A natural system of conservancy was in vogue, almost every hill top is dedicated to some local deity and the trees on or about the spot are regarded with great respect so that nobody dare touch them.

There is also a general impression among the people that everyone cutting a tree should plant another in its place.... Grass and fodder reserves are maintained, and even nap [cultivable] lands are covered with trees, wherever, though in few cases, such land could be spared from the paramount demand of cultivation. Special care is also taken by the villagers to plant and preserve trees on the edges of their fields. From this analysis, the solution logically offered was to give back to the peasants the woodland that they traditionally regarded as being within their village boundaries.

"If the village areas are restored to the villagers, the causes of conflict and antagonism between the forest policy and the villagers will disappear, and a harmony and identity of interests will take the place of the distrust, and the villager will begin to protect the forests even if such protection involves some sacrifice or physical discomfort." Pant envisaged that these areas would be under the control of the village panchayats, or councils, under whose direction the "natural system of conservancy" would once again come to the fore. As he shrewdly observed, "some restrictions will be there, but these will proceed from within, and will not be imposed from without."

Clearly, Pant drew upon and systematized the knowledge, perceptions, and analysis of the peasant folk of the Kumaun Himalaya. Thus, after the popular protests in 1921, the Government of the United Provinces set up a Kumaun Forest Grievances Committee. This committee toured the hills, examining some five thousand witnesses in all.

Peasant activists submitted dozens of petitions to the committee, on behalf of individual villages. These identified blocks of forest near every village, where peasants would have exclusive rights of fuel and fodder collection, timber for building, wood for ploughs, bamboos for basket making, etc. It was being proposed that villagers should have full rights over these forests, which they would manage through their own panchayat. Based on the evidence it collected, the committee finally concluded that "any attempt to strictly enforce these [forest] rules would lead to riots and bloodshed."

It thus divided the existing reserved forests into two categories-Class I, which were to be managed not by the forest officials but by the civil administration (in theory more sympathetic to rural needs), and Class II, constituting the commercially valuable wooded areas, which were to remain with the Forest Department. It also recommended that the government consider the constitution of village forests as per the demands of the people of Garhwal and Kumaun.

Bureaucracies move at their own pace, and only in 1930 the rules were passed allowing for the formation of van panchayats, or village forests, in the hill districts of the United Provinces. These allowed for a forest patch to be handed over to a village if it lay within its settlement boundaries, and if more than one-third of its residents had applied for permission to the deputy commissioner (DC). Once the DC gave the go-ahead, then the villagers elected, by voice vote, a council (panch) of five to nine members. This council in turn elected a head (sarpanch) among themselves. The van panchayat was empowered to close the forest for grazing, regulate cutting of branches and collection of fuel, and organize the distribution of forest produce.

PERMISSION OF THE FOREST DEPARTMENT

It could appoint a watchman, whose salary would be paid by villagers' contributions to the panchayat. The panchayat could levy fines, although if the offender did not pay it had then to go to the civil courts for redressal.

The felling of trees, however, required the permission of the Forest Department. The department also claimed 40 percent of the revenue from any commercial exploitation. Of the rest, 20 percent would go to the zilla parishad (district council), with the balance 40 percent kept with the DC on behalf of the van panchayat, which with that official's written consent could use the funds for roads, schools, and other local improvements.

There are now in excess of 4,000 van panchayats in Kumaun and Garhwal, covering an area of just less than half a million hectares. An official report of 1960 remarked that many of these village forest councils had done "exemplary work in connection with forest protection and development." A more recent survey has concluded that the panchayat forests are often in a better condition than the reserved forests. Of twenty-one panchayats surveyed in three districts, the forest stock in thirteen of these were in good condition, in four in medium condition, in three in poor condition. The researcher concluded that van panchayats have, by and large, maintained oak forests very well, especially in contrast to the dismal condition of the reserves (except for those reserves distant from habitations). The position in respect of chir forests is not so clear, but these seem to have done about as badly under van panchayat control as in the reserves. Various studies suggest that, overall, panchayat forests seem to be in as good or better condition than the reserves.

The van panchayat system constitutes the only network of village forests mandated by law in all of India.61 The concession was made by a colonial state worried of losing control in a sensitive and strategically important border region, and it was not to be replicated elsewhere. After independence, the van panchayat regime was not extended to the adjoining region of Tehri Garhwal, where Mira Behn worked in the 1950s, and where it might have very well contributed to preserving and enhancing the oak forests. Within Kumaun, too, there is considerable resentment over the curbs placed on the autonomous functioning of van panchayats. Though technically under the control of the villagers, the Forest Department can veto schemes for improvement, while of the revenue generated, 40 percent is swallowed by the state exchequer. Forty percent of the rest is by law granted to the village, but this money too first finds its way into a "consolidated fund" controlled by the DC, to which individual panchayats have then to apply.

There are signs of an emerging movement to do away with these constricting rules, to make the management of the panchayats come fully under the control of the villagers. A chronicler of this discontent, himself quite aware of the long history of forest-related protests in Kumaun and Garhwal, writes that "those who know the history of forest struggles say that... the van panchayat movement will be the biggest such movement in the hills."

TWO CHEERS FOR JOINT FOREST MANAGEMENT

From the inception of state forestry in India, perceptive critics have argued for a democratization of resource control, for a correction of the commercial bias promoted by successive governments, and for a proper participation in management and decision making by local user groups. Arguments first offered in the 1870s, and reiterated in subsequent decades, were revived, or reinvented, in the 1970s by the now-famous Chipko movement. It is no accident that Chipko originated in Garhwal and Kumaun, the part of India that has seen some of the most intensive conflicts between the state and the peasantry over forest resources.

The 1970s were marked by a series of forest movements in different parts of India. These took place in the Himalaya, in the Western Ghats, and, above all, in the vast tribal belt extending across the heart of peninsular India. In the Chotanagpur plateau, forest protests formed an integral part of the larger movement for a separate tribal homeland of Jharkhand, carved out from the huge, unwieldy, and predominantly non-tribal state of Bihar. In one much celebrated case, tribals demolished a plantation of teak, a highly prized furniture wood, that was coming up on land previously under the sal tree (Shorea robusta), a species of far greater benefit to the local economy.

Their slogan, "Sal means Jharkhand, sagwan (teak) means Bihar" was a one-sentence critique of the narrow commercial ends of state forestry. Since the 1970s, there has been an ongoing, nationwide debate on forest policy in India, a debate fuelled by the continuing social tension in forest areas and the evidence of massive deforestation provided by satellite imagery.

This debate has passed through three distinct if chronologically overlapping phases. The first phase might be designated the "politics of blame." The activists speaking on behalf of disadvantaged groups have held the forest officials responsible for environmental degradation and popular discontent. The officials, in turn, have insisted that growing human and cattle populations are the prime reason why fully half of the 23 percent of India legally designated as "forest" was without tree cover.

The forestry debate of the 1970s and the 1980s drew, at times, on the heritage of earlier movements and critiques. The peasants of Garhwal and Kumaun, as this writer found out while doing field work there in 1982-83, were acutely conscious of how Chipko itself drew on a long and honorable history of peasant resistance to state forestry. Tribal activists in Madhya Pradesh and Bihar, meanwhile, were not unfamiliar with the work and message of Verrier Elwin. And in the villages of the Deccan, social workers liked to offer the same quote of Jotirau Phule's reproduced earlier in the paper, as proof that in the agro-pastoral system of that region, proper access to forests and pasture was vital to survival, and that it was the "great superstructure" of the Forest Department that continued to deny herders and farmers this access.

However, perhaps the most direct connection between the past and the present of forest management was effected in the summer of 1982, when the Government of India circulated a new draft forest act. Activists and academics joined hands to demonstrate how the proposed legislation was solidly based upon and, indeed, took further forward the centralizing thrust and punitive orientation of the notorious Indian Forest Act of 1878. After a countrywide campaign, the draft bill was finally dropped by the state.

As tempers cooled and polemic exhausted itself, a second phase, the "politics of negotiation," originated. In villages and state capitals, forest officers and their critics found themselves at the same table, talking and beginning to appreciate, if not fully understand, the other's point of view. Concessions were made by each side, protests suspended by one, and leases of forest produce to industry cancelled by the other. One product of the growing dialogue between activists and bureaucrats was the approval, by the Indian Parliament in 1988, of a new National Forest Policy. Where the ruling Forest Policy of 1952 had stressed state control and industrial exploitation, the new document instead emphasized the imperatives of ecological stability and peoples' needs.

FARMER PARTICIPATORY RESEARCH METHODS

In Malawi and Zimbabwe, ICRISAT is developing new research methods for improving soil fertility. The Institute is also building partnerships among national scientists, extension advisors in NGOs and the public sector and farmers. The aim is to improve the ability of national research programmes to develop 'best bet' NRM technologies for poor farmers by obtaining their input at earlier stages. New technologies are needed that improve human nutrition while enhancing soil management and enabling communities to rehabilitate degraded environments. The main innovations introduced so far are intensification with long-duration and indeterminate legumes and integrated use of organics and inorganics. In Malawi, farmers are testing and adapting options such as maize grown in rotation with a doubled up legume intercrop of pigeonpea and another grain legume. Combining small amounts of fertilizer with manure and pigeonpea residues is also being tested.

APPROACH

A novel aspect of the programme is its evaluation of several participatory approaches applied in parallel in different villages. The results are compared with baseline data from villages having no known relationship with researchers or farm advisors from NGOs or extension services. This approach enables researchers and farm advisors to address their concern that farmer adoption of fertilizer and integrated nutrient management has been practically nil, despite a decade of on-farm research and the recent focus on participatory research and extension as well as training-for-transformation empowerment approaches.

Costs, as well as benefits, will be assessed for different methods of partnering. Some researchers and senior extension staff are also concerned that extension rarely reaches female-headed households and women farmers, nor do their concerns enter into agronomic research.

The methods being compared include farmer empowerment approaches led by NGOs, extension-led demonstration trials and farmer participatory research, each conducted in a different village in the case study areas. The study is determining which of the approaches are best at building institutional linkages and improving the peer relationship among stakeholders, for different locations. All of the partners involved evaluate the effectiveness and costs of each approach. Researchers are also determining how well each approach addresses the needs of female-headed households. Project partners conducted comprehensive surveys to provide a baseline for, and developed methods of, comparison. They agree that the comparison should indicate which methods are working best, as reflected in perceived cost-effectiveness and the satisfaction of researchers, extension advisors and farmers. Farmer adoption and adaptation of technologies, farmer empowerment and improved soil management will be assessed for each method.

The 'mother–baby' trial design is one method for improving communication between farmers and researchers that has so far proved successful.

The approach was originally created to facilitate farmer collaboration in testing soil fertility technologies. Further, it was conceived as a practical means for researchers to rigorously incorporate farmer evaluations of technologies at every step in the development process. The approach links together two trial types: a replicated one that fits researchers' needs and a simpler trial that meets farmer needs, rather than attempting to compromise and meet all objectives in a single trial.

Researchers first design 'best-bet' technologies, attempting to take into account farmers' priorities and resources. Then mother–baby trials are planted in each participating village. The 'mother' is a replicated experiment designed by a researcher. Farmers plant and manage the baby trials, which are single replicates of the mother trial. For this purpose, each farmer selects a best-bet technology from the mother trial and adjusts the level of inputs and equipment according to his or her preferences. Farmer evaluation is documented through surveys, community discussion and ranking exercises, which facilitate researcher incorporation of farmer input. The trials have led to improved spontaneous experimentation among farmers and give researchers and extension advisors the chance to observe and learn from farmers.

Reflections

The larger context of the different case study areas where the approaches are being tested is difficult to assess. For example, the market opportunities and historical extension–farmer relationships may vary markedly among the areas and determine the relative success of different methods. The trial and demonstration approach probably has the least emphasis on farmer knowledge, while facilitation of farmer learning and experimentation, the farmer-led approach, has the most. The changes in the researchers and extension staff to encompass a broader, more participatory approach is proving difficult to document.

Further steps, such as how to facilitate farmer training and communication with other farmers on guidelines, are not explicitly part of any approach although this may develop. Farmer-to-farmer communication is probably the most important means of technology dissemination and is not treated explicitly in any of the approaches, except perhaps the farmer empowerment approach (which some professionals view as too costly for large-scale work). This point is difficult to evaluate but should become clearer with time. A survey is underway to evaluate researcher and extension attitudes and beliefs regarding effective ways to communicate with farmers and work together to develop soil management options and to improve farmer experimentation. Surveys documenting farmer assessment of the process and farmer adoption are also on-going.

All of the partners involved in the methods' comparison systematically evaluate methodology approaches and efforts to facilitate farmer–researcher-extension linkages and technology best-bet options. Also, the concerns of women farmers and female-headed households are specifically addressed. This took time to build, in part because almost all of the researchers and extension staff involved were men and no gender sensitivity training or discussion was attempted at initial stakeholder meetings.

Attention to including women farmers in the technology development process was almost nil at first but has increased over time, particularly at the Malawi sites.

This adds value to on-going efforts in the area. The mother–baby trial and farmer empowerment approaches attempt to facilitate farmers' learning basic research principles, to expose farmers to a range of new options and to empower them to value their own knowledge. This appears to have improved communication among farmers, researchers and extension staff.

A conundrum is that extension staff and researchers mostly see the trial demonstration approach as the only cost-effective way to scale up dissemination.

Yet this approach does not facilitate farmer experimentation or joint learning. Staff at NGOs also state that their work on farmer and community empowerment is only cost-effective in isolated areas. They believe that they need to go to trials and demonstrations to reach more people, although both groups express frustration with the lack of effectiveness of demonstrations led by extension or farm advisors.

The goal of this case study is to measure long-term impact and changes in how research and extension staff conduct their work. Stakeholders designed this village-based comparison of methods and are involved in evaluating the pros and cons of each participatory method over time. Meetings are held annually for assessment. Strong links are built because researchers, extension staff and NGOs were already carrying out most of this work and through the case study we are attempting to facilitate self-reflection on the value of different approaches to the stakeholders. In Malawi, 400 farmers are assessing best-bet technologies at seven sites around the country through baby trials and their own experimentation. In the process, they are satisfying researchers' need for sound quantitative experiment results.

SOCIAL FORESTRY

Social forestry in its broad sense is a programme of community forestry development and should be characterized by the following.

1. Involvement of the beneficiary from the planning to consumption stage,
2. Production and distribution of common forest products suck as fuel, fodder, small timber, fruits and other things which were obtained earlier from the forest,
3. use of community lands to produce such products,
4. adoption mixed production system *i.e.* grass, fodder, fruits and other things people require,
5. Maintenance of minimal government control in production, management and distribution of benefits,
6. Collaboration by the government, panchayat and voluntary agencies in the supply of inputs (funds, seedlings, materials and labour),
7. Measures taken to conserve soil and water,
8. An in built mechanism for recycling of organic matter and waste of the area,
9. Full use of all the natural resources in social forestry area,
10. Training components to make people to reflect both on the social aspects as well as technical aspects,
11. Educational measures to bring villagers together as a community based on the egalitarian values and
12. Encouraging people to help themselves not only to survive but also to develop.

SCOPE OF SOCIAL FORESTRY

In India 35% land remains unused or underutilized. This includes cultivable waste lands, permanent pastures and grazing land, land under miscellaneous tree crops, follow lands, barren and uncultivable lands and lands along road sides, railway lines, canal banks and drains. Most of these wasteland can be used for social forestry programmes. The catchment area of village percolation taken is an excellent place for social forestry plantation. Seasonal crops should not be cultivated in the catchment areas to avoid soil erosion and siltation. But implementation of a planned social forestry on the same area increases infiltration of rainwater, reduces siltation to minimum and augments water storage in the reservoir. Social forestry on the catchment area and its percolation tank from an ideal eco-system with a symbiotic relationship between them. The former increases water storage in the tank and avoids siltation whereas the latter supplies moisture to trees during dry seasons.

The demand for fuel wood in India is steadily increasing. The estimated current firewood consumption is 157 million tones. Most of it is used for cooking. To cook 1 kg of food, 1.2 kg of fuel wood is required. This clearly indicates the need for producing more fuel wood than food. About 400 million tones of cattle dung, which is equivalent to about 60 million tones of fuel wood, is burnt annually in India. The annual loss of plant nutrients by this wasteful practice amounts to approximately 300 thousand tonne of nitrogen, 156 thousand tones of phosphorous and an equal amount of potassium. If cattle dung were incorporated into the soil instead of using it as fuel, it would increase food grain production by about 15-20 million tones. This can be achieved only when fuel wood is supplied to the rural poor in sufficient quantity.

Fodder problem of livestock population needs immediate attention. The fodder available at present is very low both in quality as well as in quantity. It can meet only 23% of the protein requirement and 39% of the carbohydrate requirement of the livestock population. Growing perennial fodder crops becomes a necessity to solve the fodder problem. Similarly, seeds and fruits of some trees are excellent ingredients for concentrates. All these can be grown in social forestry. Small timber and bamboo are usually used by the poor for building and repairing their houses and for making agricultural implements, as cement, iron etc. are beyond their purchasing capacity. The demand for small timber and bamboo will remain for a long period and will increase along with the increase in population. Social forestry can provide the rural poor with timber and bamboo. Similarly, social forestry can solve the food problem of the poor to a great extent. Certain edible fruits like cashew, mango, coconut and palms have high nutritional value and can grown under social forestry programmes. Social forestry can play a vital role in the reclamation of degraded lands, conservation of soil and moisture, improvement of agricultural production and prevention of environmental deterioration.

PRINCIPLES OF PLANT SELECTION

Selection species is important for the success of social forestry. The selection should be based on the following criteria:

1. trees should be fast growing, early maturing and yielding,
2. they should have multiple usages (for food, fodder, fuel, manures),
3. the tree trunk should be strong and stout,
4. the species should be suited to climate and soil of the place,
5. they should have dense foliage,
6. they should possess capacity to tolerate adverse climate and soil conditions,
7. they should have deep root system,
8. the leaf shedding should be in early spring and not in summer,
9. they should have coppicing capacity (ability to produce shoots from cut portion),
10. they should not have prominent thorns and
11. their planting and caring should be easier and economical.

It is essential that the species selected for planting should be directly or indirectly related to the soci0-economic needs of the people. Therefore, there must be full participation of local people in selecting species for planting. At the same time people have to be taken through the process of reflection and analysis to make them aware of the socio-economic conditions of the their life in relation to social forestry before they choose the type of trees to be planted. Planning should take into account not only the immediate needs but also the future increasing requirements.

PLANT GROUPS

Trees can be grouped according to people's requirements. For the selection of trees, people should identify locally available species first and then only go for exotic species. This principle should be kept in mind always before a species is selected for social forestry. One can always find at least half a dozen local species suitable to various needs of the people. Given below is a model classification of trees according to the needs of people and major agro-climatic conditions. However, the list is neither exhaustive nor the classification rigid or exclusive. The scientific names with Hindi names in parenthesis are given in the classification. It is up to the people to identify local trees for various purposes and plant them.

Fuel Trees

Fast growth and high yield are the most important criteria for selecting firewood trees. Tress requiring 5-15 years to attain full growth are considered suitable for fuel production. Fuel wood can be grown by two methods: (a) coppicing (cutting and regrowing branches continuously). All the trees do not have regrowth capacity. Some fast growing cropping fuel trees are: alder eucalyptus, subabul, poplar and sallow. Some of the common firewood trees are classified climate wise though some of them are grown under a wide range of climatic conditions.

Tropical dry regions: (a) *Acacia nilitica* (Babul); (b) *Albizia lebbeck* (Siris) (c) *Azadirachta indica* (Neem); (d) Emblica officinalis (Amla); (e) Prosopis julifera (Vilayati babul); (f) Zizyphus mauritiana (Ber); and (g) Tamarindus indica (Tamarind).

Tropical humid regions: (a) *Acacia nilitica* (Babul); (b) *Albizia lebbeck* (Siris) (c) *Azadirachta indica* (Neem); (d) *Casuarina equisitifolia* (janglijhau); (e) *Dalbergia sisoo* (Shisam); (f) *Ficus religiosa* (Pipal); (g) Salix daphnoides (Willow); (h) Madhuca longifolia (Mahua); (i) *Moringa olifera* (Drumstick); (j) *Tamarindus idica* (Tamarind); (k) *Leucaena leucocephala* (Subabul); and (1) *Mangnifera indica* (Mango).

Subtropical regions: (a) *Acacia catechu* (Khair); (b) *Grevillea robusta* (Silver oak); (c) *Populaus spp*. (Popular); (d) *Salix spp*. (willow); and (e) *Toona ciliata* (Toon).

Fodder Species

Fodder species should be included in a social forestry programme to ensure balanced feed to the animals.

Fast growing leguminous species are more valuable as fodder due to their high protein content(Subabul). In addition to being a legume, trees with succulent and dense foliage with better silage making quality are preferred for fodder. Fodder trees are generally planted close and maintained in the form of a bush by repeated coppicing. Some of the fodder trees are classified according to climate.

Tropical dry regions: (a) *Acacia nilotica* (Babul); (b) *Leucaena leucocephala* (Subabul); (c) *Cordia rothii* (Gondi); (d) *Zizyphus mauritiana* (Ber); (e) *Albizia lebbeck* (Siris); (f) *Bauhinia vareigata* (Kachnar).

1. *Tropical humid areas:* (a) *Morus alba* (Shatoot); (b) *Moringa olifera* (Drumstick); (c) *Acacia nilotica* (Babul); (d) *Albizia lebbeck* (Siris); and (e) *Artocarpus heterophyllus* (Kat-hal).
2. *Subtropical regions*: (a) *Ficus religiosa* (Pipal); (b) *Toona ciliata* (Toona); (c) *Bauhnia purpurea* (Lal kachnar, Kaliar); and (d) *Acacia catechu* (Khair).

Fruit Trees

Choice of the local people should be given due consideration for selecting fruit trees in social forestry along with soil and climatic factors as the fruit trees do not florish unless they get a suitable environment. The objective for raising the fruit trees should be also ascertained. If it is for commercial purpose, marketing facilities should be first ensured. Following species can be grown according to the specific region.

1. Tropical dry regions: (a) *Zysyphus mauritana* (Ber); (b) Tamarindus indica (Tamrind); (c) *Moringa olifera* (Drum stick); (d) *Emblica officinalis* (Amla); (e) *Carissa carandas* (Karaunda); and (f) *Ceratonia siliqua* (Kharnub).
2. Tropical humid regions: (a) *Anacardium occidentale* (Cashew); (b) *Areca catechu* (Supari, Betelnut); (c) *Artocarpus heterophyllus* (Kathal); (d) *Cocos nucifera* (Coconut); (e) *Tamindus indica* (Tamerind)l (f) *Madhuca longifolia* (Mahua); (g) *Embnlica officinalis* (Amla); (h) *Mangifera indica* (Mango); (i) *Moringa olifera* (Drumstick); (j) *Morus Alba* (Shahtoot); and (k) *Syzygium cumini* (Jamun, Jambu).
3. Subtropical regions: (a) *Litchi chinensis* (Litchi); (b) *Emblica officinalis* (Amla); (c) *Magnifera indica* (Mango); (d) *Psidium guajava* (Guava); (e) *Prunus persica* (peach); and (f) *Pyrus communis* (Pear)

Timber Trees

Timber trees are of two types namely small timber and commercial timber for logging. For social forestry purposes, normally small timber is raised which is used for village housing for making farm implemts and tools handles. Here, too, the local species are given preference over the exotics. Some of the major timber speciaes according to their purpose, climatic requirements are grouped.

1. Rural housing and fencing: (a) *Albizia lebbeck* (Siris); (b) *Azadirachta indica* (Neem); (c) *Cocos nucifera* (Coconut); (d) *Madhuca longifolia* (Mahua); (e) *Syzygium cumini* (Jamun); (f) *Dalbergia sisoo* (Shisham); and (g) *Phoenix dactylifera* (Khajur).
2. Big housing and furnitures: (a) *Acacia catechu* (Khair); (b) *Acasia nilitica* (Babul); (c) *Cassia fistula* (Amaltas); (d) *Dalbergia sisoo* (Shisham); (e) *Madhuca longifolia* (Mahua); (f) *Pinus spp*. (Pine); (g) *Shorea robusta* (Sal); (h) *Tectona grandis* (Teak); and (i) *Terminalia arjuna* (Arjun, Kahu).
3. Farm implements: (a) *Acacia catechu* (Khair); (b) *Acacia nilotica* (Babul); (c) *Azadiracht indica* (Neem); (d) *Cassia fistula* (Amaltas); and (e) *Syzygium cumini* (Jamun)
4. Tool handles: (a) *Pongamia glabra* (Karanj); (b) *Leucaena leucocephala* (Subalul); (c) *Bambusa spp*. (Bamboo); (d) *Grewia oppositifolia* (Bhimal); (e) *Azadirachta indica* (Neem); (f) *Acacia catechu* (Khair); (g) *Acasia Nilotica* (Babul) and (h) *Dalbergia sisoo* (Shisham)

Trees for soil erosion control: for soil erosion control, planting methods and designs are more important than the kinds of tree. Generally trees useful for soil erosion control are classified.

1. Tropical dry regions: (a) *Acacia catechu* (Khair); (b) *Acacia nilotica* (Babul); (c) *Albizia lebbeck* (Siris); (d) *Azadirachta indica* (Neem);

(e) *Bambusa spp.* (Bamboo); (f) *Leucaena leucocephala* (Subabul); and (g) *Prosopis julifera* (Vilayati babul).

2. Tropical humid regions: (a) *Acacia catechu* (Khair); (b) *Acacia nilotica* (Babul); (c) *Albizcia lebbeck* (Siris); (d) *Azadirachta indica* (Neem); (e) *Bambusa sp.* (Bamboo); (f) *Leucaena leucocephala* (Subabul); (g) *Prosopis julifera* (Vilayati babul); (h) *Salix sp.* (Willow); and (i) *Pongamia glabra* (Karanj)
3. Subtropical regions: (a) *Populaus spp.* (Poplar); (b) *Eucalyptus spp*, (Eucalyptus); (c) *Bombax ceiba* (Semal); (d) *Azadirachta indica* (Neem); (e) *Dalbergia sisoo* (Shisham); (f) *Morus alba* (Mulberry); and (g) *Prosopis julifera* (Vilayati babul).

Trees with the following characteristics are suited for soil erosion control:

1. profuse branching from the base (eg.Vilayati babul),
2. extensive root system (eg. Casurina)
3. trailing branches (eg. Cashew),
4. ability to produce saplings from cutting (eg. Shisham); and
5. tolerance to close growing condition (eg. Bamboo).

Improvement of Poor Soils

Saline, alkaline, *usar*, waterlogged and marshy land constitute the poor soil. These lands cannot be used for agricultural purposes. Planting trees resistant or tolerating these types of soils, sill not only help in reclamation of soil, but also yield useful products like fuel or fodder. Planting trees on water logged areas will lower the water-table to certain extent. Following trees are suitable for the improvement of soils.

1. Saline, alkaline and usar soils: (a) *Acasia nilotica* (Babul); (b) *Albizia lebbeck* (Siris); (c) *Azadirachta indica* (Neem); (d) *Butea monosperma* (Dhak); (e) *Casurina spp.* (f) *Emblica officinalis* (Amla); (g) *Madhuca latifolia* (Mahua); (h) *Morus alba* (Shahtoot); (i) *Pongamia glabra* (Karanj); (j) *Terminalia arjuna* (Arjun), (k) *Zizyphus spp.* (Ber); (i) *Prosopis julifera* (Vilayati kikar); (m) *Nerium indicum* (Kaner); (n) *Cassia fistula* (Amaltas); and (o) *Ficus religiosa* (Pipal).
2. Short duration water logged areas: (a) *Acacia nilotica* (Babul); (b) *Butea monosperma* (Dhak); (c) *Dalbergia Sissoo* (Shisham); (d) *eucalyptus spp:* (e) *Pongamia glabra* (Karanj); (f) *Terminalia arjuna* (Arjun); (g) *Syzigium cumini* (Jamun); and (h) *Salix spp.* (Willow).
3. Damp and marshy lands: (a) *Pongamia glabra* (Karanj); (b) *Salix spp.* (Willow); (c) *Popular spp.* (Popular); (d) *Casuarina equesitifolia* (Jangli Jhau); and (e) *Toona ciliata* (Toona).

Windbreaks

Windbreak refers to one or three rows of trees planted as a protective against strong wind. Windbreaks are planted perpendicular to the direction of the wind in the locality. In India, there are generally planted north-south direction to check the wind force from south-west and north-east. They reduce the wind velocity and consequent movement of sand and soil is arrested.. thus they (1) reduce soil erosion (2) conserve precipitation and (3) increase soil moisture. The following are a few examples of trees suitable for planting as windbreaks.

1. Dry and arid regions: (a) *Albizia lebbeck* (Siris); (b) *Azadirachta indica* (Neem); (c) *Artocarpus heterophyllus* (Kat-hal); (d) *Pongamia glabra* (Karanj); (e) *Sesbania grandiflora* (Agast); (f) *Tamarindus indica* (Tamarind) and (g) *Tamarix dioica* (Jhau).
2. Coastal areas: (a) *Anacardium occidentale* (Cashew); (b) *Grevilea robusta* (Silver oak); (c) *Casuarina equisitifolia* (Casurina), *JangliJhau)*; (d) *Saraca indica* (Ashok).

CONSTRAINTS OF SOCIAL FORESTRY

Several factors constrain social forestry programmes. Some of the more significant ones are mentioned below:

A. Absence of comprehensive research findings on the suitability of certain species for various agro-climatic regions of the country, techniques for growing such plants successfully and economically, optimum spacing of trees, subsequent management and harvesting practices etc. is a major hurdle for social forestry.

B. A remunerative support price for tree corps grown by beneficiaries is lacking. Instead the middlemen take away most of the profit.

C. Rules and regulations regarding the harvesting, transport and marketing have to be simplified so as to eliminate hurdles in obtaining permission. For example, sandalwood production in Karnataka has been steadily decreasing despite the increasing demand. One reason for low production is absolute state control over sandal trees even on private lands. Similarly most of the trees except eucalyptus, babool, saijana and few others are forbidden to be cut and transported outside the village even by the producer himself without permission from the forest department. The farmer, there, is reluctant to grow these trees as he has no right either to cut or sell them.

D. Harassment by government officials is another problem in the implementation of social forestry programmes. They demand illegal gratification from villagers for transporting and trading in minor forest products produced under social forestry programmes. Hence people are not interested in social forestry programmes.

E. Most social forestry programmes are undertaken on community wastelands, roadside and canal margins where it is very difficult to grow and protect the saplings. Moreover, most of the plantations are very small in size and scattered making management and protection impossible.

F. Encroachments of the land also deprive the poor from his due benefits. The rich and the influential people encroach and take possession of most of social forestry plantations and the poor are marginalized. People should be involved at every stage with real conviction and understanding. Otherwise they become mere workforce. Special task forces need to be mobilized in every region to create awareness among the people and to motivate them. It should be made a movement both in thinking as well in practice. For the success of social forestry it must become a people's movement involving as many villagers as possible.

G. Extension is a two-way communication between researches and people; better and new ideas about social forestry should flow from researchers to people and the people should provide the feedback to the researchers. At present there are neither such facilities nor competent personnel in extension methodology.

H. Lack of proper distribution system for the sharing of the social forestry products is yet another constraint. Neither government nor voluntary agencies have evolved an acceptable and equitable system of product distribution.

I. Non-availability of required seedlings is another major handicap. Hence, people plant whatever they get which ultimately proves unsuitable to them.

SOCIAL FORESTRY INTERVENTION

Social forestry is any practice, method, technique/technology or natural resource management system that enhances forest resource governance and makes forestry economically viable, and ecologically sound. This definition is mostly a state view of what social forestry should be, and does not emphasise the involvement of local people. It advocates that forest plantations should be controlled by the state within fences, thus excluding local people and livestock from entering it. Projects designed with such an approach, for example the National Social Forestry Project (NSFP), aided by the World Bank in India, are typical examples dominated by the orthodox views of deforestation. In general, social forestry interventions initially emerged from concerns over forest degradation. These forestry interventions, whether it was community forestry (CF) in Nepal or social forestry in India and Sri Lanka, had their focus on technical questions of resource management. It was only towards the beginning of 1990s that the shortcomings of the approaches became evident and since then social forestry interventions started taking a new direction. The forestry interventions started to encompass a rhetoric on a more 'people-oriented approach'. Social forestry or CF was then redefined as a concept, "where forest is managed by local people (living near or in the concerned forest) for their own benefit". However, the new approach still does not explicitly address livelihoods and local resource management patterns. The fundamental concept of CF in Nepal is that local forest users are given collective management responsibility for the local forests on which they depend for product flows. The model is an extremely defined form of organization that is defined in relation to the resource and not to existing social relations or community structures. The new policies still emphasize more on protection and regeneration of the resource, in response to the heightened concerns over rapidly deteriorating forest resources.

Under the new management approach, the Joint Forest Management (JFM) committees in India or the Forest User Groups (FUGs) in Nepal find it more difficult to operate in some situations, as the state agencies become more strict as these are supported by policy amendments. For example, the Forest (Second Amendments) Bill, 2001 in Nepal amended the 1993 Forest Act in a number of fundamental ways, which have restricted the FUGs. Under this amendment, the FUG revenues will be taxed at 65 per cent, and CF in the *Terai* area would be subject to separate and more stringent regulations. This demonstrates the dual nature of the state trying to promote local participation on one hand, and at the same time giving more powers to itself to retain control of the forestland.

In general, justifying social forestry intervention was much easier for government agencies in situations where deforestation crisis was projected as more serious, than in situations where it was considered less serious. In the latter case, it can be difficult for forest agencies to justify their intervention and secure funding. On the other hand, the cost-benefit ratio had to be projected to convince donors about the economic viability of the projects. This led to the recommendation of fast growing species in social forestry plantations, irrespective of local factors and needs. Eucalyptus was one such species recommended by forest planners on a large scale during the 70s and 80s in social forestry projects. Large areas were brought under monocultures of Eucalyptus, which became a major environmental controversy towards the 80s and 90s. In the Terai region of Nepal and India, the Forest agencies, replanted degraded *sal* (Shorea Robusta) forests with exotic tree species, mainly *Eucalyptus* and *Sissoo* (Dalbergia Sissoo), under the pretext that it would produce biomass in short periods. There was resistance from local NGOs opposing the replanting programme of the Forest Department, in some cases leading to violent conflicts with local people.

Social Forestry has been part of the national forest policies in South Asian region in one form or another since the 70s. But after three decades of development and implementation, it has still not met its stated objective of promoting plantation activities, which address the needs of poor people for fuel, fodder, small timber and non-timber forest products. According to Agarwal (1995), social forestry has the potential for substantial welfare improvements for the target population, especially women, through increased biomass availability, decreased time for collection and decreased pressure on natural forests. Intervention needs, however, to be very selective to be successful; they must be locally specific with special consideration given to the needs of local people. The goal of social forestry intervention should not only be to provide biomass but also to reverse the processes of negative change through active participation.

NATIONAL POLICIES AND SOCIAL FORESTRY

The Forest Policy prescriptions during colonial and neo-colonial periods in South Asia have been influenced by the mainstream views on deforestation leading to restrictions on local communities to extract biomass from state forests in this region. Most of these policies continued in the post-colonial period, but under a different nomenclature. International policy documents on environment and development such as WCED (1987) have also been echoing such views and providing support to national policies in developing countries. These restrictive forest policies and legislation have rather weakened the State in its efforts to conserve natural resources.

In general, forest legislation is south Asia is based on a 'command and control' system, stating mainly what is not allowed, rather than how the forests *can* be used.

The system creates numerous opportunities for corruption, including private sale of produce from the state forest, payments for overlooking offences and unofficial 'fees' for issuing permits for logging and transport. Particularly the latter two have created widespread resentment among local people, and conflicts with the forest agencies. State failures have led to the realisation by the governments and conservationists in South Asia that the orientation and direction of past state policy on forest management had itself been the primary reason for the worsening state of forests. For example, The Indian Forest Policy, 1988 mentioned that:

" Excessive concern for revenue, a disregard for people's needs, and lack of initiative in involving people in the past needs to be corrected ".

Such policy statements need to be highlighted to show that conflicts actually originate in the dissolution of local institutions and ignoring local people's needs.

Local environmental movements in south Asia have also become more influential in the present decade, especially the grass-roots environmentalism based on the legacy of issue-focused local voluntary bodies (*e.g.*, *chipko* movement in India), and the political ecology movement, which presents a wider critique of modernization. Grassroots environmentalism and political ecology have both been able to gain momentum by involving society at the local level.

Informal grassroots forest protection movements in south Asia such as the "Western Ghats campaign" and "Save the Himalayas", promoted environmental concerns as well as local interests through media and rehabilitation programmes.

The pressure from various movements across the region and international agencies eventually led to new forest policies in south Asia during the 90s. The new forest policies of Sri Lanka, India and Nepal appear to be moving towards participatory use and management of forest resources and the involvement of local people, community groups and the private sector. Also, local organisations like the FECOFUN in Nepal, have gained power though the development of a national non-government organisation.

This organisation has been able to influence political decisions, and is putting pressure on forest authorities to allow local management and local retention of revenues. Pressure is also being asserted to reduce corruption within the forest departments. Thus, the push for change in policies and institutions is building up from beneath, and supplements a push for change coming from international cooperation and changes in development policies.

This new approach to forest management attempts to promote people's participation through co-management. This approach could be a part of the new people-centred paradigm recognizing that local people are capable of managing their environment. The new approaches offer a promising opportunity to respond to the numerous problems which states are facing while implementing forestry projects. However, the approaches are also co-opted by the state and, given the scepticism in the administration to give power to the people, the implementation of the new policies and the participatory role of local people the new paradigms is not clear. Is it simply to co-opt rural people into state-dominated modes of resource use in order to lessen state obligations toward resource management and enforcement, or will local communities really be involved in decision-making?

According to some, the new forestry management paradigms are genuine attempts to involve local people in natural resource management. There are others who are sceptical about it, however, since it has been framed externally in a top-down manner. The top-down nature is evident from the policy documents where the state defines the rules and terms of partnership in forest development projects. The local communities are meant to follow these terms, which do not show much flexibility. For example, the ratio of benefit sharing between forest protection committees and Forest Departments in general are fixed by the forest agencies. Activities such as tree felling and fodder extraction are possible only with the prior approval of these agencies. This also applies to plantations created under social forestry projects on forest as well as community lands.

The major constraint in implementing the new forest policies lies in the hesitation of the State bureaucracy to transfer powers to local communities.

This is evident from the presence of the local forest officer who represents the state in the village committees constituted under new forest management models. The local forest officer will not only be a technical advisor, but will also control permits for use and sanctions for misuse of natural resources. The state has rights of exclusion here, since it decides who may or may not have access to forest resources. The village committees do not have any powers to frame rules of withdrawal or impose sanctions on offenders. The most common approach to resolve offences so far was through paying bribes. Thus the system of permits controlled by state officials was vulnerable to corruption and manipulation, and its continued existence only served to highlight the vulnerability of the state authority. The Forest Service personnel see the new decentralized policy as a threat to their authority. Many problems associated with the new decentralized approach stem from the policy guidelines, which are interpreted to suit the interests of the Forest Service

According to Kothari *et.al.*, (1995), externally initiated community forest management efforts in South Asia have not proved very effective in controlling deforestation problems. With recent socio-economic changes in the region the present situation with reference to biomass extraction may not remain the same in the future if resources are not managed properly. Markets may start influencing forest management in the region encouraging people to cut trees. It all will depend on how the new policy initiatives will actively empower local people and to what extent the people will be involved in management of natural resources.

State initiatives to help local communities in meeting their energy demands through village woodlots or closure schemes have remained as isolated attempts. In a few areas, state forests and village lands that were closed periodically under rehabilitation programmes by Forest Departments were found to have good vegetative cover. This may be due to additional protection measures which the department provides to closures, but in general such measures were not popular since they often excluded local people, as seen from the damage to plantations, frequent grazing and low success rate.

SOCIAL PARTICIPATION AND FOREST POLICY

Decentralization in Bolivia was motivated by increasing pressure from civic committees seeking to have greater control over natural resources, the general trend towards decentralization in neighbouring countries and the prominence of decentralization in donors' agendas. Its implementation has been affected by social participation dynamics, and the new forestry regulations have had implications for democratization of forest resources access.

ANTECEDENTS OF DECENTRALIZATION

Bolivia has three levels of government: central government; departmental government or prefecture, whose main authority (the *prefecto*) is appointed by the president; and municipal government, in which democratically elected municipal councils elect the mayor. Before decentralization most decisions were made at the central level. The municipalities had limited resources and little influence in policy-related decisions, even those directly affecting the development of their municipal jurisdictions. The Bolivian lowland forest area was largely marginalized from the political centre in La Paz, where most political decisions were made. In the early 1960s that region was progressively integrated into the national economy through the expansion of natural gas extraction, agriculture and logging. Those factors fostered the growth of a regional elite, expanded the contribution of that region to the national income and increased its influence on development programmes. In the 1970s, the government took its first step towards decentralization by establishing corporations for regional development. This was closer to an administrative attempt to transfer some investment decisions to the departmental level but the central government still appointed the presidents of such entities, and most of the decisions were negotiated at higher levels. Civic committees (groups of local social organizations) had been seeking greater access to forest revenues and greater participation in the formulation of forest policies since the mid-1970s. In the 1980s, legislation was approved establishing the collection of a forest fee of 11 per cent, to be used in regional development projects. (In 1993, further legislation was passed stating that forest companies would pay 80 per cent of their taxes directly in the areas where the resources originated, but the mechanism was hard to implement in practice.) By the late 1980s, the national forestry service was deconcentrated to local branches. This move towards deconcentration, however, did not make the forestry service a more efficient and less corrupt institution. Municipal authorities, meanwhile, continued to respond to the leaders of their respective political parties rather than to their constituents.

Decentralization was prompted through approval of Popular Participation Law (No. 1551) and Administrative Decentralization Law (No. 1654), both passed by Congress in 1994. The first altered the responsibilities of the municipal governments, while the second modified the responsibilities of prefectures or departmental governments. In 1996 a new Agrarian Reform Law was issued, as well as a new Forestry Law. The Agrarian Reform Law's objective was to define the legal basis for a system of titling and land regularization, and to redefine the conditions of access to rural property. The Forestry Law attempted to redefine the conditions for obtaining and maintaining forest rights. Both affected the way in which landholders and forest users can access and maintain their rights for forest resources use.

PARTICIPATION IN INVESTMENT DECISIONS

The Popular Participation Law expanded the municipal government's jurisdiction beyond the urban centers to the whole rural area within the municipal borders. It made municipalities responsible for local schools, health facilities, roads maintenance and water systems. To finance these new responsibilities, the central government allocated 20 per cent of the national budget to the municipal governments, to be distributed among municipalities in proportion to their populations. Both rural and urban property taxes were earmarked for the municipal governments, who now administer their collection. The Popular Participation Law has strengthened municipal governments and made them more democratic. Rural populations – mainly smallholders and indigenous people – have gained the right to participate in municipal elections and run for the municipal councils. Nonetheless, national political parties still appoint individual candidates, allowing political parties to maintain their control over local political agendas, and to reproduce a system of political patronage with local leaders. Local candidates who want to run for office have to negotiate their agendas within the priorities of the political parties, and if they are elected they are accountable to those parties.

Furthermore, the law sought to introduce community control over municipal governments by local social organizations (ie local farmer organizations, neighbourhood committees and indigenous groups) and community-based vigilance committees. Nevertheless, these committees' representatives are exposed to continuous pressures from political parties.

Rules and Regulations for Forest Use

The Forestry Law of 1996 defined a set of regulations for forest use, somewhat differentiated according to forest user, under the premise that sustainable forest management is feasible under the right practices. It established a new system for monitoring forest management, enforcement and sanctions to illegal logging, as well as introducing some market-oriented regulations and taxes to discourage unsustainable forestry operations. The goal is to achieve sustainability of forest management through progressive incorporation of less valuable timber species and the application of extraction techniques that promote natural regeneration. Furthermore, the law seeks to create clear rights over forest resources, thereby encouraging investments in forest management; and to eliminate forest crime and illegal logging, as well as set technical criteria for forest management. The public institutional system of the forest sector was substantially altered. The Ministry of Sustainable Development and Planning is the ruling entity, the Forestry Superintendence is the regulation entity, and the National Forestry Development Fund is the financial entity.

Noncommercial forest uses do not require authorization, and a forest management plan is an essential requirement for all types of commercial forest activities. Hence, forest concessionaires as well as private landholders are obligated to design management plans as an instrument to regulate commercial logging activities, including forest inventories and mapping. Forest management plans have to comply with many technical requirements. Forest management, when based on selective management, must respect a minimum cycle of 20 years between logging operations on the same area, and a minimum cut diameter must be respected. Furthermore, annual operations plans are required. The regime to be applied to nontimber forest products is similar.

Democratizing Access to Forest Resources

The new forestry regulations included two provisions that have had some impact in democratizing access to forest resources. The first refers to the exclusive right of indigenous peoples to use the forest resources within their indigenous territories, recognized legally by the Agrarian Reform Law. According to this law, indigenous claims are considered titled after completion of a process of land regularization. Currently, a total of 19 million ha have been claimed for titling as indigenous territories, but it is not known how much will ultimately be effectively titled in this way. Stocks (1999) estimated 5 million ha to be the area with commercial logging potential in areas claimed as indigenous territories. A provision of the Forestry Law states that local forest user groups can benefit from forest concessions within areas declared as municipal forest reserves, which represent up to 20 per cent of public forest within each municipal jurisdiction. Local forest users can be granted a forest concession if they are recognized as a local user association by the Ministry of Sustainable Development and Planning. This mechanism was conceived as a way to formalize the access for local forest users or small-scale loggers who, because they had no legal right to access forest resources, were previously conducting forestry operations informally.

POWERS TRANSFERRED TO MUNICIPAL GOVERNMENTS

According to Ribot (2001), outcomes of decentralization depend on the type of powers that are transferred to lower levels and on who receives such powers. In Bolivia, as well as in most countries of Latin America, decentralization has followed a top-down format, and municipalities have been the main recipients of authority transferred from the central level.

A Top-Down Model of Delegation

Although the Popular Participation Law did not grant municipal governments any new explicit function related to natural resources management, it motivated some municipal governments, as a result of their larger political authority, to become involved in natural resources issues and thereby capture part of the benefits. Because mayors became relatively more powerful, the central government and the international donors began to consider them more seriously as partners in environmental projects. The increasing political power of municipalities led the government to consider them for dealing with some problematic issues, such as social monitoring of illegal logging and formalizing forest rights for small-scale loggers and other local users. These were included in the Forestry Law of 1996.

The current powers and functions are divided as follows:

- Formulate forest policies, strategies and norms;
- Determine land classification and evaluate forest potential;
- Prepare demarcation of concession areas for timber companies and local groups;
- Set prices for concession fees and volume-based taxes;
- Promote research, extension and education; and
- Solicit technical assistance and funding for forestry projects.

Forest superintendence (SF):

- Supervise technical compliance with the forestry regime;
- Grant management rights to eligible forest users;
- Approve forest management plans for different forest rights;
- Enforce forest regulations and sanction illegal forest users;
- Issue concessions, authorizations and logging permits;
- Request external forest audits of forest operations; and
- Collect concession fees and volume-based taxes and distribute them.

Prefectures:

- Formulate forest development departmental plans;
- Develop forest research and extension programmes;
- Promote programmes of rehabilitation of degraded forest systems; and
- Develop programmes for strengthening municipal forestry units' institutional capacities.

Municipal governments:

- Propose the delimitation of municipal forest reserves up to 20% of available public forest;
- Protect and conserve the reserve areas until they are conceded to local user associations;
- Inspect and control all forest activities within their territorial jurisdiction;
- Report violations of forest regulations to the Forestry Superintendence;
- Provide support to local forest users in implementing their management plans;
- Establish the registry of forest resources in their jurisdiction;
- Develop soil use plans corresponding to the departmental use plans;
- Organize training events for local forest users; and
- Facilitate and promote social participation in local forest development.

To carry out their new responsibilities, municipal governments are expected to create municipal forestry units. Municipalities can form consortiums with other municipalities to create such units. In theory, the entire system should be entirely financed with the revenues coming from both concession and clearcutting fees.

Prefectures receive 35 per cent of the concession fees and 25 per cent of the fees charged for clearcutting operations. Municipal governments get 25 per cent of both types of fees. The National Forestry Development Fund receives 10 per cent of the concession fees and 50 per cent of the clearcutting fees. SF gets 30 per cent of the concession fees. A reduction of forest concession fees in March 2003 has diminished the financial resources going to prefectures and municipalities.

Restricted Powers for Municipalities

Municipal governments must comply with national regulations regarding property regimes and forest regulations. Hence, their autonomy to make decisions about natural resources depends largely on decisions made at the central level. The legislation thus regards municipal officials as rule followers rather than rule makers, and as implementing agencies for policies defined at the central level. In short, the new institutional forest system has not delegated important responsibilities to municipalities and has therefore not led to dramatic changes. The central level reserves for itself major decisions regarding allocation of forest resources rights, granting forest concessions, approving forest use regulations, and collecting taxes from forest resources use. Municipalities have little to say in these areas, and all informal actions of municipalities for collecting forest taxes and controlling timber transit are considered illegal.

Although municipal governments are empowered to control forest crime, promote community forestry and take other forest-related actions, their power to make autonomous decisions regarding the allocation and use of forest resources is restricted. For instance, they can decide how to allocate the forest resources within the municipal reserves but cannot decide about the size of such reserves. This reflects the tension between the central level, which defends the strong role of the national forestry service, and those who support a more active and autonomous role for municipal governments.

Limited Autonomy for Local Forest Users

The situation of indigenous groups has improved. Indigenous communities whose forest resources were subject to encroachment from illegal loggers now have exclusive access to resources within their territories, though problems persist. However, before indigenous people can take commercial advantage of forest resources, they have to develop forest management plans according to

Forest Superintendence regulations, and they have little scope to adapt such norms to their own management practices (BOLFOR, MDSMA, and Superintendencia Forestal, 1997).

And even though specific regulations for forest management in indigenous areas have been approved, local knowledge has been ignored in much of the forest policy. Enforcement of indigenous property rights depends on some institutional arrangements made at the departmental and regional levels, over which indigenous people have little influence. This issue is linked to a more complex bureaucratic process of land use planning and titling with poor overall outcomes.

The other social actors who have benefited from access to forest resources – small-scale timber extractors and other local forest users – also have no autonomy to make decisions about the way in which they use the resources, and all of them have to comply with forestry regulations to maintain access to public forest resources through forest concession systems.

The Forest Superintendence has instituted command-and-control mechanisms to enforce the implementation of what it considers good forest management practices among local forest user groups and small landholders. Those practices, though appropriate for large-scale forest concessionaires, act as barriers for small-scale forest users.

Implementation of Decentralization Policies

There are always failures in the implementation of policies, and decentralization is no exception. In Bolivia, transferring responsibilities to municipalities meant building local capacities in municipal governments, as well as interacting with actors who had disparate incentives and interests to use the forest within the new institutional context.

Building Institutional Capacities

Municipalities with forest resources began to receive their shares of forest taxes in 1997, when they started to set up municipal forestry units. The resources transferred to municipalities have tended to decrease because forest concessionaires did not comply with the forest fees payments. In 2002, a new system formally reduced the amount of taxes collected by the state from forest operations to almost half of what was originally expected, formalizing the reduction of transfers from the central government to municipalities. By the end of 2001, almost all municipalities with forest resources (about 109) had created their municipal forestry units, and each had at least one forest or agricultural technician. In municipalities with difficult access and little population, the forest units are the only local providers of technical services, and governments tend to value their role in local planning. The main limitations are the staffs' lack of technical skills and inadequate training in social issues, such as conflict resolution.

Building technical capacities in municipal forestry units constituted an important step forward in relation to the past. Nevertheless, progress is not uniform and varies with the financial support the units receive from the municipalities. That in turn depends on the amount collected in forest fees and what proportion of it goes to supporting operational budgets. A significant portion of municipalities consider it enough to provide minimum resources to the municipal forestry units and spend the rest, if any, on other activities. It is worth mentioning that local government priorities are providing social services and infrastructure, and they have little motivation to support productive projects , though this is changing.

Incentives for the Actors

The new institutional system resulting from both decentralization and the shift in forest policies has modified the incentives for municipalities and forest stakeholders to continue doing what they were used to doing, or to adapt their social and financial strategies to the new conditions. The main incentive for municipalities to engage with the new decentralized order of things was the likelihood of getting a share of the forest taxes, as well as the possibility of administering the municipal forest reserves. Municipalities also are interested in penalizing illegal clearcutting due to the fact that they get a portion of the fines, but this does not happen in relation to illegal logging. Furthermore, municipal governments have been interested in being active players on forest issues in the cases where local forest users have political influence on local decisionmaking, or in cases where forest-dependent people represent important votes to keep them in office.

The incentives of local forest users to engage in the process are diverse. Indigenous people have received important benefits from forest policy reform and decentralization. They have gained rights to make exclusive use of their forest resources within their territories, and have the chance to expand their influence to participate in municipal decision making. In this context, forest management might increase the social legitimacy of indigenous people's claims to land. In turn, local small-scale timber extractors' main incentive has been to get formal access to forest resources through the forest concession system, and to benefit more from formal markets. Furthermore, this group should benefit from the technical assistance provided by the municipal forestry units, which are receiving resources from the forest taxes to support such activities. Both groups have benefited from several forest projects (eg BOLFOR) and NGOs with an interest in supporting community forestry.

In the short run, the main losers under decentralization were absentee forest concessionaires, who must now negotiate with the forestry service. Furthermore, a forest tax has been imposed, and they have had to acknowledge indigenous people's demands. The main incentive for forest concessionaires to adapt to the new conditions was their interest in keeping their forest areas, and an implicit commitment from the central government to help them through a difficult financial situation originating in the timber crisis in regional markets. Municipal governments reacted in contradictory ways regarding forest concessions. Although some are critical of concessions' activities within their jurisdictions, others consider them sources of economic growth and employment.

Homogeneous Solutions for Heterogeneous Municipalities

The Bolivian model of decentralization does not account for regional variation, and its design assumes that all municipal governments will react uniformly to the challenges arising from the new conditions. Reality proved rather different.

Implementation of decentralization has had to face three issues: uneven distribution of resources to municipalities, the minimal availability of public forest declared as municipal forest reserves, and the hetereogeneous interests of municipalities regarding forest-related activities.

Uneven distribution and allocation of financial resources. The financial resources allocated by the Forestry Superintendence to municipalities have varied. From 1997 to 1999, only 30 municipalities benefited, with more than 80 per cent of the total resources transferred to them. In the richer municipalities, not all income from forest taxes is spent on forest-related activities; some is diverted to other sectors. The poorest municipalities likewise have urgent demands that are outside the forestry sector.

Minimal municipal forest reserves. Even though the law specified that up to 20 per cent of public forest would be declared municipal forest reserves, in practice such areas were not available in all municipalities – in some cases because of the existence of overlapping claims over public forest, in others because public lands were maintained as forest concessions. In some municipalities there were no areas to declare as municipal reserves to allocate as concessions to small-scale loggers. By early 2001, from a total of 2.4 million ha demanded by municipalities as municipal reserves, only 681,000 ha had been designated.

Municipalities' priorities. Only a few municipal forestry units have accomplished the functions they have been granted. Municipalities prioritize their investment in response to social pressures or to local authorities' political motivations. The level of municipal engagement is also related to municipalities' incentives to respond to their constituents and be accountable to higher authorities. For instance, whereas some municipalities are interested in allocating resources to establish a system of forest concessions within municipal reserves, others choose to control illegal clearcutting or make forest management viable for small farmers.

Uneven Outcomes

Transferring responsibilities to municipalities has modified the political and institutional arrangements for forest management, with implications for both people and forests.

Participation in Local Politics

In many lowland municipalities, small farmers, indigenous people and small-scale loggers have been elected to office for the first time. Indigenous groups have been able to obtain political support from the municipal councils to reinforce their land claims, small-scale loggers have obtained support to negotiate temporary logging authorizations, and small farmers have obtained support favouring their efforts to modify land-use and forest regulations. Thus, municipal governments have contributed to political support for some local actors' claims on resources. This is also the case where these groups have strong organizations that can influence the municipality's decisions or where they represent the majority of voters. In these cases, municipal governments may amplify the demands of social actors. In other cases, however, transferring responsibilities and resources to municipalities has reinforced the power of preexisting local elites – elite capture, particularly in municipalities of northern Bolivia, and where cattle ranchers and timber companies are highly influential in local politics. These local elites have, in some cases, influenced municipal governments to build local alliances against indigenous land claims and strengthened their power over use of resources. The local elites of these municipalities often benefit from activities based on mining of natural resources and promote concentration of benefits in a few hands. Decentralization, in these cases, tends to produce undesired results in social equity and forest conservation.

Where the social composition of local government is more complex, local elites have to negotiate with small farmers and indigenous people. It is not unusual to find alliances built among the different social groups that support certain development agendas, such as infrastructure development and basic social services. Those social agreements are, however, difficult to arrange for natural resources management, though in some cases alliances have been established to protect conservation areas against encroachment of foreign timber and mining companies (ie municipalities of Rurrenabaque and San Ignacio de Velasco).

Much of the impact of decentralization on reconfiguring social participation in local politics depends on the local political economy and on the social capital of marginalized groups. In other words, the more democratic municipalities have traditionally had more equitable access to resources. Those in which local elites have imposed their interests are affected by the extent to which local groups have been able to build social capital.

PARTICIPATION IN FOREST MANAGEMENT GROUPS

Participation in forest management groups has been shown to influence decisions to plant more trees on-farm. This, however, does not imply that households not participating in these groups are unlikely to plant. Participation in forest management groups greatly enhances people's attached value to forest ecosystems and the need to protect them; which in turn results in their desire to increase forest cover on their farms. Participation in community-based conservation groups enhances farmers' access to diversity, quality and quantity of tree species.

Groups increase farmers' awareness on importance of trees and market access to the tree and other agricultural products. The argument for market access, however, only applies where markets work fairly well. In such cases, farm households would plant high value timber and fruit trees, not only to satisfy their monetary needs but also subsistence requirements. This implies that closeness to market could also accelerate participation of households in establishing farm forests through two distinct channels—providing market for products of farm forests and creating alternative off-farm income generating avenues which compels households to set their own sources of fuel wood because of lack of time for collection from other sources far from households.

PARTICIPATION IN COMMUNITY FOREST MANAGEMENT

Besides Participation in community forest management, households' decisions to plant trees may be directly influenced by household-specific, plot-specific and institutional factors. For instance, farm forests have enormous environmental advantages beyond direct benefits to the farm households. To comprehend these indirect benefits, the decision-maker at household level requires some education, either formal or informal, obtained through schooling or extension services.

Thus, better educated household heads or households with access to government or farmer-farmer extension services are better adopters of farm forestry, either because they view tree planting as a means of improving the land or because they are able to appreciate other non-quantifiable benefits as ambiance, micro-climate modification or carbon sequestration. This also explains why households with good social networks may have a higher possibility of planting trees because they are able to get extension services through such networks.

Households to Plant Trees

Institutional factors have also been shown to influence the decision by households to plant trees. Secure land tenure arrangements, for example, have been found to influence tree planting decisions among farmer groups. Trees take a longer gestation period and only farmers who are confident of continued use of a given plot would be encouraged to plant them. However, some studies do not agree with the idea that secure tenure may encourage tree planting and cite cases where communal ownership of land has been more conducive for development of farm forestry. Perhaps tree planting in areas with ambiguous land tenure system is a means used by households to place a claim of legitimacy of ownership and/or access.

CONCEPTUAL FRAMEWORK

Kakamega forest region on which this study is based lies in the rural parts of Western Kenya. Like other parts of developing countries, agricultural households are expected to contend with either missing or highly imperfect markets characterized by high transaction costs and constraints on marketed quantities. Specifically, only a few households buy/sell forest products. Equally, very few households hire labour for farm activities or extraction of forest products from the neighbouring government forest. As a result, this study is premised on the non-separability of household production and consumption decisions first introduced by Singh *et al* (1986) and advanced by others like de Janvry *et al* (1991), Sadoulet *et al* (1996) and Taylor and Adelman (2003).

This study adopts a theoretical framework of utility maximization by a household envisaged by Singh *et al* (1986). The household derives utility from consuming market goods, home-produced goods and household leisure time.

Analytical Model

The area of study has embraced community forest management introduced by the Forest Act (2005). Individuals are free to choose whether to join Community Forest Associations or not. However, only members of Community Forest Associations have the privilege to access a specified range of forest products. Members of Community Forest Associations, because of their access to forest resources from various parts of the larger Kakamega forest, may have no drive to establish own farm forests or they could simply reduce the number of trees planted on own farms. Alternatively, these people could expand forest cover on their own farms due to increased appreciation of forests and the need to protect them. This self-selection choice made by households which want to belong to Community Forest Associations (CFAs), if ignored in the analysis, would lead to inefficient parameter estimates and biased intercept. The argument, here, is that the unobservable household factors that drive participation in CFAs could be related to those that drive household tree planting behaviour. Thus, correcting for selection bias is a precursor to our analysis of level of farm forestry establishment by households in the area of study.

The study hypothesizes that besides conventional productive factors, individual, household and institutional factors contribute to farm forestry development at various levels of resource endowments. These factors express themselves through correlation of self-selection of households into CFAs or not and level of farm forestry development, and through inclusion of observable factors in explaining inter-household variation in levels of farm forestry development. The main econometric task is to determine how self-selection and other factors influence levels of farm forestry among farm households in Kakamega.

Following Heckman (1979) and Flood and Grasjo (1998; 2001), we specify our model, ordered into two regimes. The first regime is defined by whether a household is a member of a Community Forest Association or not.

The Study Area and Data

The study site for this survey was around Kakamega Forest, situated in Kakamega District in Western Province of Kenya. It lies north-east of the Lake Victoria between latitudes of 00°10'N and 00°21'N and longitudes of 34°47'E and 34°58'E at about 1600 m above sea level. The forest area is drained by two main river systems, the Isiukhu River to the north and the Yala River to the south.

The forest is the only remaining rain forest in Kenya and is the furthest east remnant of the Guinea-Congolean rain forest. According to the 1994 welfare monitoring survey, 52% of the population in the district lives below the poverty line, meaning that they can hardly afford basic necessities like food, shelter, clothing, and education. As such there is a heavy reliance on the forest to supplement their daily necessities. This region has also been considered by the Kenya Woodfuel and Agro-forestry Programme (KWAP) as one of the areas that could benefit most from policies that target improvement of forestry projects due to its high population and high agricultural potential.

The household characteristics show that the average age of the household heads is 48 years, with the average number of schooling years being 9. The low education level of the household head is a contributor to their inability to secure more remunerative employment opportunities elsewhere, thereby resorting to farming activities. Education increases household's off-farm employment opportunities.

Furthermore, highly educated members of the household tend to look for greener pastures in off-farm activities. This is because of the traditional nature of farming activities within the region which many people view as not competitively rewarding compared to non-farming activities. The average household size measured in this survey is 5 members. Of the households interviewed 73 percent were male headed.

On the farm characteristics, the average farm size in Kakamega is 2.2 acres; 76 percent of households practice farm forestry while the mean value of household assets is KES. 22026. The average distance to the nearest forest edge is 1.5 km. On average, 23 percent and 17 percent of the households had access to extension services and credit facilities, respectively. With regards to institutional attributes, 73 percent of households participate in social groups while 52 percent participate in community forest management through CFAs. In terms of tenure security, 38 percent of households owned land with title deeds while 59 percent owned non-titled land.

The Selection Equation

In the selection equation, variables that significantly and positively influence selection include the household's land size, the square of the selling market distance, education of the household head, household's awareness of the new Forest Act and security of land tenure. Those that have negative influence include distance to the government forest, distance to the selling market, and male-headed households.

Land size, being a proxy of household's wealth, could positively influence the choice to join CFAs because such households are able to pay the requisite fees for participation. Large landholders are also associated with large herds of cattle and this may increase demand for grazing fields, attainable from forests through membership to CFAs.

The convex nature of the effect of distance to selling market on the selection could be embedded in the fact that such a distance influences access to information on formation and operation of CFAs, and for those who extract forest resources with the intention of selling, longer distance to the market makes joining CFAs less attractive. Reversal of the effect of market distance on selection after some limit could be explained by the fact that markets tend to be near urban centres. Thus, very far away from these markets are most likely typical rural settings with greater needs for primary livelihood products attainable from forests. This creates the impetus to join CFAs.

Better educated household heads are likely to be aware of formation of CFAs and the potential benefits. Consequently, where the head is better educated, the household has a higher chance of participating in CFA. It is also evident that households that enjoy secure land tenure are more likely to join CFAs. This is perhaps due to the fact that such households are made of native residents, who find it easier to work together with the other residents they have known for a longer time, some even being close relatives. Such natives may also have special attachments to the government forest, either for religious or cultural reasons.

This creates the urge to join CFAs so as participate in protection of the forest and gain access to it as and when need arises. Households that are aware of the new Forest Act, which actually introduces community participation in management of forests, are more likely to join CFAs. Such people definitely understand that it is only through participation in CFAs that they stand to gain from the government forest.

Distance to the government forest influences selection negatively because one of the main reasons for participating in CFAs is to gain access to the forest for extraction of specific forest products. As a result, those who live far from the forest have no motivation to join CFAs. Similar results are associated with male-headed households most probably because such households are able to raise own farm forests from which to obtain the basic products which would have otherwise made it necessary to access government forest through CFAs.

The Outcome Equation

In the outcome equation, distance to the selling market appears to have a concave relationship although only the coefficient of distance to the market is significant at 10 percent level. The reason for this could be that those living far from the market have to rely more on subsistence production of most of their domestic requirements. If they need forest products, they may be compelled to plant their own trees. The geography of Kakamega is such that the forest is closer to the town. The implication of this is that, if town is the main market, those living far from the market are also far from the government forest, the alternative source of forest products.

Of special interest in the outcome model are the coefficients of household heads, household size, social capital and farm size. All are significant and positive; indicating that, holding other factors constant, male-headed households, larger households, larger land size and more participation in social organizations are associated with planting of more trees. On the contrary, secure land tenure is associated with planting of fewer trees or under-development of farm forestry. The fact that male household heads are associated with planting of more own farm trees is more cultural than economics. Planting of trees is viewed more as a man's activity than a woman's.

Where labour markets do not function properly or where the household is less endowed with resources to hire labour, men could be better placed to plant trees because this activity is more strenuous. Other activities like fencing and construction of household's dwelling units are also viewed as a man's responsibility and these could compel a man into planting more trees in anticipation of future needs.

Larger households tend to plant more own farm trees. There are two main angles to this—larger households have larger requirements for forest products such as fruits, fuel wood and medicinal plants. This could make it more prudent and economical for such households to establish own farm forests. The second angle is that tree planting is labour-intensive and larger household are capable of using own labour to accomplish the tasks involved. Diversification of livelihood sources to meet the pressure of feeding a large family may provide an alternative explanation.

CRITICAL FOR FARMER-FARMER EXTENSION

Social capital is critical for farmer-farmer extension and also for raising resources, physical, human and financial. Thus, households with wider interaction networks through participation in social organizations plant more trees, either because they have drawn lessons from others in their network loops or because they have benefited from rotational labour common among rural households.

Through the social networks, relatively poor households can borrow the requisite farm tools or even obtain free tree seedlings from other group members. Households with large pieces of land, all else equal, plant more trees. This is because trees compete with other crops for land. Those with smaller pieces of land may, therefore, devote the entire parcel for food crop production as opposed to large land holders who are able to either inter-crop food crops with trees or even set aside portions of land exclusively for trees. For households that use trees to mark plot boundaries, it follows that that the larger the piece of land, the more the trees planted.

The coefficient of security of land tenure looks perverse at first sight. One would expect households with secure land tenure to be more motivated to plant trees because trees take a longer time to manure. However, on second thought and taking cognizance of the Kenyan situation, trees are a means of entrenching land ownership, either on plots that are not regularly cultivated or on untitled land. Consequently, households with insecure land tenure are predisposed to plant more trees, not as an economic investment but as evidence of ownership should disputes arise.

Marginal Effects

Marginal effects computation combines effects of variables included in both the outcome and the selection equations, conditional on the dependent variable being observed or selected. After accounting for selection, coefficient of distance to the selling market is negative while the coefficient of the square of the distance to the market is positive, showing a convex relationship. This implies that the number of trees planted by households falls as distance to market increases up to an inflection point and then starts to rise for households that lie very far from markets. Possible explanation to this is that households plant trees not only to meet their domestic needs but also to generate income. As such, the households near markets plant more trees because they are motivated by the possible large profits. But very far from the market could be a reflection of typical village life where demand for fuel wood, medicinal plants, fruits and fodder is very high yet each household has to rely on own production. What was noticed during the survey is that jaggery-making is a major economic activity in the remote villages of Kakamega—this could be fanning demand for fuel wood which translates into increased planting of trees.

Secure land tenure is found to positively influence the number of trees that a household plants. The underlying explanation to this is that trees take a longer time to mature and one would invest in them only if he/she is certain of using the land for a longer time. Similarly, size of landholding by the household turns out highly significant, increasing the number of trees planted. A household with a small piece of land may devote the land for production of subsistence crops and may use the land too intensively to allow planting of trees. Conversely, households with large pieces of land are capable of either undertaking agro-forestry or setting aside substantial portions of their land for growing trees. Even if trees are confined to farm boundaries, a larger plot would still accommodate a fairly large number of trees.

Other factors that increase the chance of planting more trees are the education level of the household head and household's awareness of the new Forest Act. Better educated household heads are able to appreciate the role of forests—in soil and water conservation at the farm level and carbon sequestration at the global level. Such household heads may also be more inclined towards ensuring that the government forest remains intact, making them develop their own sources of forest products.

The Forest Act (2005) which introduces community participation in forest management gives communities increased access to the forest but only for a restricted, well-defined range of products. It is also more punitive on the violators and the fact that it involves communities increases the chance of the violators being caught. Thus, for those who understand this Act, it would be more prudent to develop own forest if the interest goes beyond products allowable from government forests.

For variables that appear only in the outcome equation, the parameter estimate is indeed interpreted as the marginal effect. Therefore, social capital can be said to be positively related with the number of trees planted by a household, other factors remaining unchanged. This is attributable to flow of information among members of social groups on the positives of planting trees. Other contributions of social capital could include inter-household rotational use of labour and farm implements which is well developed among rural households. Farm forestry is labour-intensive may require households to pool their labour resources especially in instances where households are not well endowed with financial resources to hire labour or where labour markets are not well developed.

This study focuses on how participation in community forest management, introduced in Kenya by the Kenya Forest Act (2005), and other factors impacts on development of farm forestry which has been perceived as a vehicle through which to increase the country's forest cover. Using household data collected from Kakamega forest communities and controlling for selection bias, the study find makes key findings with various policy implications.

First, households that are involved in community forest management through Community Forest Associations (CFAs) are associated with under-developed farm forestry possibly because they are able to meet their demands for forest products from government forests. The policy implication of this is that, while the new system of forest management may be important for protection of existing forests, it could be counter-productive to increasing the total forest cover in the country. An intensive campaign for farm forestry development should therefore accompany it.

Secondly, secure land tenure is important for development of farm forestry because trees take a longer time to mature. It is therefore important to guarantee security of access or ownership of land in the country. Properly functioning land market could be one of the options of ensuring that rights of land owners and land users are equally protected, and that longer tenancy periods are enforceable. Thus, the government should work towards developing land markets.

Third, small landholders are hesitant to plant trees possibly because the land cannot accommodate both crops and trees. One thing that peasants should be made to understand is that investing in trees could be more profitable than investing in food crops. Proper education and introduction of high value fast maturing tree species could overturn the current perception of the peasants.

Fourth, education level of the household head and his/her awareness of the Forest Act (2005) favour development of farm forestry. This underscores the need to target household heads in dissemination of information on importance of forests at household, national and global level, and the content of the current Forest Act.

Lastly, social capital among households is useful in increasing forest cover. As a result, it is important to develop infrastructures that promote formation and operation of social groups. For instance, the government could provide tree seedlings or farm forestry funding through community groups.

PRODUCTIVE CAPACITY OF HOUSEHOLD FOREST MANAGEMENT

In India, Joint Forest Management (JFM) is one of the recognized organizational systems to regenerate forest resources, meeting local needs of people and sharing expected benefits. In order to conserve, regenerate and enrich forest resource base as well as to reduce the dependency of forest dwellers and people residing on the fringes of forest, the Government of Tripura passed its own resolution to establish an organization called Joint Forest Management Committee (JFMC). The main objective of this study is to examine the role of JFMC on Household's productive expenditure in Tripura State. The study attempts to analyze the trends of the productive and non productive expenditure incurred by JFM committee members as well as non-JFM committee members in Tripura State. In Tripura, more than 70 percent of the total population is engaged in agriculture and allied activities like animal husbandry, forestry, and fishery. It contributes about 43 percent of the net state domestic product. However high density of population coupled with high growth rate and rapidly declining landholding, has led to decline in the productivity of agriculture. Furthermore, lack of non renewable resources coupled with inadequate infrastructure has created hurdles in the path of growth and development of the state. There has been hardly any significant development in the industrial sector. As a result, the local population is increasingly becoming more dependent on forests to supplement their income.

Also, it is a paradox that where more forests exist, there is higher degree of poverty and to mitigate this problem several policy initiatives have been tried but still much is required to be done. Several questions arise concerning implementation of JFM on a sustained basis. So it becomes necessary to study the role of JFMC in enhancing the productive capacity of people of Tripura State.

SOCIO-ECONOMIC CONDITION OF FOREST MANAGEMENT

The key hypothesis considered here is that the socio-economic condition of JFM Committee members is better than that of non-JFM committee members. The study suggests some valuable policy inputs to execute the JFM institutional framework more efficiently. It concludes that the socio-economic condition of people associated with JFMC in South district has improved than North district. The paper has been divided into four parts – part one begins with the review of literature. Part two highlights the characteristics of the study area, data collection and descriptive statistics of the variables. Part three highlights the econometric model adopted to test the hypothesis and finally part four provides results with strong policy implications based on the findings of the study.

PROMOTION OF SUSTAINABLE FOREST REGENERATION

There are several studies on the role of JFMC in different parts of the world. The study conducted by Kumar (2002) at Jharkhand concludes that JFM is well suited to the promotion of sustainable forest regeneration, but such regeneration has been achieved at the expense of the poor. In the absence of significant levels of inter-group mobility and thus of reduction in the level of inequality between poor and non poor villagers, it is imperative that actions should be taken to put in place compensatory mechanisms for rural poor in order to secure a proper measure of equity. The present study considers the inequality through incorporating the variables namely income from agriculture and allied activities, income from forestry, income from livestock etc. Adhikari *et al* (2003) concludes poor households face more restricted access to community forest resources than less poor or relatively better off households. Their analysis suggests that collection of forest products from community forest is dependent on various socio-economic variables. The variables like landholding, livestock, caste, education and households' economic status exert a strong influence on appropriating benefits from the commons. Adhikari (2005) in his study on Nepal's common forestry found that income from common forestry resources is related to socio-economic attributes and private endowment of households. The households with land and livestock assets and also belonging to upper caste gain more income from commons forestry resources. Thus the less poor households are, in absolute terms, more dependent but in relative terms, less dependent on forest resources than the poor. Also, the female-headed households benefit less from common forests, thus, further aggravating the inequity in distribution of benefits.

The present study considers the income distribution of poor and richer people to evaluate evaluates their expenditure on productive and non-productive activities. Behera (2006) states that community elites continue to dominate actual decision-making process despite efforts by policy makers to empower poor and weaker section people in JFM area. According to Bandyopadhyay *et al* (2004) village proximity to forest, leadership and fuel wood dependence are significant factors in influencing their participation in common forestry.

It has been found that participating households consume more fuelwood than non-participant households. Their analysis reflects that fuelwood consumption is closely tied to fuelwood collection and decrease with increase in assets. It also includes education as one of variables which could influence their participation in the JFM committee meeting. Lise (2000) concludes that a high dependence on the forest and good forest quality enhances voluntary people's participation. Low average level of education in family and high levels of education of respondents and greater involvement of women in the community stimulates participation. In addition, state has a major role in stimulating the people's participation. The present study explains gender related issues through inclusion of gender variable in the econometric model. Dhanagare (2000) highlighted the weakness in the institution and lack of clarity about the sharing of usufructs among beneficiaries and between forest department and the people. The author stressed on the partnership between forest department and the people for the success of JFM.

DETAILS OF THE PRIMARY SURVEY

The Profile of Study Area

The study focuses on the South Tripura and North Tripura where the forest department functions through two circles – northern circle and the southern circle. Out of the total 6292.68 sq. km. of Government Forests in the state, 3093.96 sq. km. lies in the northern circle and 3198.72 sq. km lies in southern circle. The forests in the South Tripura District are administered through three territorial divisions, namely Udaipur, Bagafa, Gumti and one Wildlife Sanctuary at Trishna. The Conservator of Forests based at Udaipur monitors the activities of the Divisional Forest Officers (DFOs)/ Wildlife Warden in South Tripura district and West Tripura district. In South Tripura, 182 JFMC/EDCs are functioning covering a project area of 36054 ha including 17527 hectares of land covered under plantation. The total membership of JFMC/EDCs in the district is 12045. The community wise classifications show that 54.29% are STs, 22.48% are SCs, 14.61% are from general category which mainly includes the upper caste groups and remaining 8.62% are people belonging to OBC group.

The study has chosen Simsima as a main study area in South Tripura. Simsima is a panchayat village located in development block of Matabari, Tehsil Gorjee. The village falls in the circle of district South Tripura. The village has a JFMC named 'Azad Hind Baanaayan Committee'. The effort to form this committee was initiated in 1996 but it was only in 2002 that the committee was recognized by the forest department. The memorandum of understanding (MoU) was signed between the forest department and the committee on 11thJuly, 2003. During the initial days the people were hostile to the view of formation of committee as it viewed the committee as an agent of government to prevent them from cutting trees for fuelwood. However, the JFMC was formed and gradually the objectives of the committee were brought to the notice of the community through organizing awareness campaign with the assistance of forest department.

This formed the motivation for choosing Udaipur in South Tripura as a study area. The close relationship between the people and forest; the philosophy of 'protect and prosper' brought the people especially tribals closer to JFMC. The tribal 'Garo community' also started to participate in the affairs of the JFMC. However, there is no representation of tribals in the executive committee. There are seven members in the executive committee; the executive committee is dominated by the members of the OBC community. There is only one woman in the executive committee as against the two (33%) required mandatory according to the guidelines.

There is no ST member in the executive committee. All the members of the executive committee are educated. The annual membership fee of the committee was Rs.20/- only. The committee has opened a joint account in the Grameen Bank, Chanderpur, which is operated jointly by the Member secretary and the treasurer of the committee. Because of these issues, the study includes the variable like social groups and savings of households. The practice of shifting cultivation (*jhuming*) has reduced with the formation of the JFMC. The members of the committee are vigilant about the illicit felling of the trees and the offender is caught and is handed over to the forest department. In primary study, 62 percent respondents are members of the JFMC in South Tripura. The general body meeting is held once in year in the month of March-April. However no general body meeting was conducted in the 2006-2007. Initially during 2002 the JFMC was provided with the land of 80 hectare, in 2005 it was further provided with 120 hectare land. It now has 200 hectare land. The JFMC has taken over plantation of hybrid *Amla* and Bamboo, the benefits of which shall be obtained after ten years. The wage employment for the members of JFMC is meager, only two days employment was provided last year to the members during the seed sowing of hybrid *Amla* and Bamboo. The wage employment is insufficient as there is no vacant land in the forest area belonging to JFMC.

This motivated us to explore the income generation from the forest activities. The vital issue in the village is the prevalence of poverty and economic insecurity which forces people to work hard and for long hours in the fields. About 40% families are either land less or do not have sufficient resources to take up farming on their own land. The fact that people are always engaged in economic activities for meeting their daily needs leaves little time for them to participate in the activities like JFMC. The JFMC has been obtaining funds from the forest department for investing in the income generation activities of the members. The fund flow to the JFMC is not regular which acts as bottleneck in the planning activities. Finally, it further motivated us to analyze the income from the agriculture, horticulture and livestock.

The forests in the North Tripura district are administered through four territorial divisions, namely Kanchanpur, North, Manu, and Ambassa. The Conservator of Forests based at Kumarghat monitors the activities of the Divisional Forest Officers (DFOs). Chinibagan is the para located in Deoracherra. Deoracherra is an Autonomous District Council (ADC) village that falls in the development block and tehsil of Gaur nagar in Kailasahar sub-division of district North Tripura. The village has a JFMC named "Briksha Mitra". The MoU of the committee was signed in July 2002. The committee is not registered with the forest department. Majority of people inhabiting the project area live below poverty line and work as casual labour. In addition to that the other economic activities prevalent in the project area are collection of minor forest produce such as bamboo, thatch, firewood etc. The project area has mixed community comprising of Hindus, Muslims and Christian. The committee now has a land of 350 hectares available for plantation and other NTFP activities.

Data Sources

The required data for the study was obtained through a primary survey of 200 households using detailed face-to-face questionnaire/ interview schedule and secondary data. The survey was done in two phases. Firstly, a pre-pilot survey was done. In this, the areas were visited initially and the problems were analyzed. After the pre-pilot survey, two districts were selected namely North

Tripura and South Tripura representing and supporting the objective of the study. From the each district one representative village was selected namely Chinibagan in North Tripura and Simsima in South Tripura districts, based on the purpose of the study. The target villages were selected on the following criteria, accessibility from road, having more than 250 households, degraded forest (crown density below 40%), dependency on forest resources for livelihood, presence of JFMC. Before conducting the main survey a pre-testing of the interview schedule was done at village Tulakona in West Tripura. The questionnaire was tested on 10 households and was finally modified accordingly. The main household survey was conducted for 10 days with the assistance of ten enumerators having sociology background consisting of 6 boys and 4 girls, fluent in Kokborok, Bengali, Hindi and English languages. The data was collected from 100 households selected randomly in each village on their demographic details, socio-economic condition (level of education, social group, income from different sources, land and assets holding), their expenditure pattern, their dependency on forest and lastly about their awareness of JFM and its benefits.

DETAILS ABOUT SURVEYED QUESTIONNAIRE

The surveyed questionnaire contains a brief introduction about the study and is divided into seven sections. The villages (name, gram shabha, block, sub-division, district) and the respondents (name, age, social group, religion, gender, education, main occupation and supplementary occupation). The covers socioeconomic data of the family (age structure, family composition, level of education, occupation etc). The detailed information about the annual income (income from agriculture and allied activities- crops, horticulture, livestock, forestry products and income from non agriculture and allied activities), yearly expenditure (productive and non productive expenditure) and savings. The covers details on assets ownership (land holding, livestock possession, housing specifications-size and type, household assets, agriculture assets, vehicles possession etc). The households' dependency on the forest for fodder, fuelwood and other non-timber forest products and contains information on JFM (participation of people in the JFM committee meetings, people's perception about the JFM committee's work plans, types of benefits derived, etc). The questionnaire was prepared in two languages English and Bengali.

4

Community-based Forest Management

When, after decades of dispossession and disempowerment, shifting government policies make it possible for a devolution process to take place, local people's organizations can only rejoice. But assuming full responsibility for the environment they live in requires that communities acquire new skills and new capacities, while building their social capital. Although not a panacea, participatory development communication (PDC) can greatly help communities to take up the new challenges associated with decision-making and self-determination.

The Bayagong Association for Community Development Inc (BACDI) is an upland people's organization in Aritao, Nueva Vizcaya, the Philippines.

Its members were victims of the involuntary relocation project brought about by the government's construction of a dam that submerged the people's ancestral lands. Instead of taking the government's offer of resettlement in a place not akin to their culture, they migrated to a nearby forested land. This experience precipitated their engagement in some kind of participatory development communication as they needed, then, to discuss their fate and how they could cope with it. Of course, these people did not know that they were involved in PDC!

Labelled as squatters and encroachers by government, they struggled for almost 28 years to hold on to the lands that they have been *de facto* occupying since they migrated in 1960. Periodically during the past, they had to resist and endure the efforts of government agents and threats from outside speculators to expel them from the area. To make their appeals known to the concerned authorities, these upland farmers engaged in dialogues with government officials and authorities. But when their voice seemed not to be heard, they resorted to street protests and lobbying. These strategies were outcomes of their frequent discussions and analysis of their situation. A wind of change worked in BACDI's favour when the government's policy on natural resource management adopted community-based forest management as the national strategy for sustainable development of open forestlands. BACDI applied for and was awarded the Community Forest Stewardship Agreement, a tenurial instrument that formally legitimized their claim over the forestland they have been occupying. It granted them tenure of 25 years, renewable for another 25 years.

While the political and social intent of such a devolution strategy was highly appreciated, issues emerged regarding the community's readiness and absorptive capacity to handle the devolved tasks and functions. Devolution as used here refers to 'the process where the locus of power and control shifts from the state to the local communities'. Under community-based forest management and as part of their new set of responsibilities, BACDI members underwent the process of participatory resource management planning. Here, PDC took the form of a social preparation method, equipping the people with the knowledge and skills that they would need as forest resource managers. Through community mapping, resource identification and problem prioritization, they learned to draft their plan and called it their own.

The other gender-sensitive participatory rapid appraisal methodologies used for situational analysis included community resource profiling; gender-disaggregated household activity and decision-making; stakeholder identification and analysis; political-ecological mapping; historical-structural analysis; institutional analysis; participatory analysis of problems and options; and resource management action planning. In all these methods, PDC was a core process for learning and planning.

Supplementing this variety of PDC methods were communication tools such as Venn diagrams, maps, community billboards and posters, written documents of their organization, and policies agreed upon for managing their resources.

This exercise paved the way for the community members to have a better grasp of the quality of their resources, their cultural integrity, their social capital and their political capacity, as well as a sense of their prejudices, weaknesses and vulnerabilities – knowledge that they could have not unravelled had they not undergone PDC. This also enabled them to identify internal and external threats to their resource management and how these might be handled.

To a large extent, the participatory experience enabled them to implement their plan with more confidence and better direction. Whereas before, the community simply remained buried in their culture of silence and subservience, the knowledge acquired of their rights and responsibilities through community meetings and discussions enabled them to become more open and assertive. As a result, when they conducted social mobilization among their members and allies, they were already more aware of community organizing, government policies, their internal capacities, external possibilities and the given constraints in the field.

Participatory development communication paved the way for BACDI's link-up with other government and non-governmental organizations (NGOs).

Members knew that they could not muster all the resources needed to manage their forest resources well. They eventually realized that they also needed to access external assistance, such as livelihood opportunities, *barangay* roads, markets, credit, schools, health centres and other social services. Partnerships became another imperative, and they were able to establish alliances by using PDC.

As they learned more about themselves and their resources from the various participatory rural appraisal methods used in the study, BACDI members were able to come up with more rational plans and approaches for managing their communal forest. In the past, they had left these matters to their leader and whoever was deemed influential in their group. PDC also enabled them to reorient their outlook and address their needs by first using locally available resources before turning to outside sources for assistance.

For monitoring and evaluation, they employed informal, unstructured discussions where they would ask each other and reflect upon where they were at a certain time relative to their plans and targets. The observations they made were then discussed in their regular community meetings. Here, members were allowed to clarify and validate their observations.

Used in the various facets of participatory resource planning, PDC has actually served as a mechanism and, at the same time, as the context of group learning. As they progressed in learning about their biophysical, socio-cultural, economic and political environments, BACDI members became more enlightened and rational managers of their forest resource. Even though most of them had only attained a low level of education, and some were even unschooled, they felt that they had learned many things from their community natural resource management activities. As a result, they were able to delineate the bounds and limits of the physical and political space that they had claimed for themselves. This was reflected by the sample plan and set of policies formulated governing the sustainable use and protection of their resources.

Social dialogue became a frequent activity in which they engaged by virtue of the communal nature of their forest resource. Decisions had to be collective and inclusive. Although discussions were not always smooth sailing, the members of the community have gradually learned the techniques of negotiation and consensus-building. In the process, their dwindling social capital has been enhanced. Social capital refers to cooperative social relations and collective action processes. This is embedded in 'norms or reciprocity, networks of civic engagement, trust, and obligations that facilitate coordinated activities' (Asian Development Bank, 1994). Since social capital depreciates through time, constant dialogue and open communication prevented it from deteriorating. Strong social capital has enabled the BACDI community to protect their traditional resource system from outside encroachers and the centralizing tendency of the state.

As BACDI members learned to master the science and art of forest management, they were continuously bombarded with challenges. Dominant among these was the multi-stakeholder setting for decision-making and action planning. It was through PDC that community members were able to respect and manage the diversity of views on an issue. Constant dialogue opened them to many possibilities and made them realize that there could, indeed, be many possible solutions to a problem. Likewise, PDC provided the venue for mainstreaming women in a men-dominated BACDI ethnic group. From mere cooks and servers during community meetings, women started assuming more substantive roles, such as being a liaison officer, secretary, treasurer and even *barangay* councillor. It must be emphasized that it was not PDC alone that paved the way to local forest management by BACDI members. Other factors were also involved, such as forest culture, social capital, policy presence and assistance by external actors. It can be safely assumed, however, that participatory development communication played a critical role, tempering the socio-political environment, internal and external to the community, so that a climate favourable to the community's takeover of resource management was created.

REFLECTIONS

Just like any other communication methods, PDC is not a panacea. It also has certain caveats, nuances and limitations. As a tool, its users and practitioners have to understand its profound intricacies. It takes time and practice, and a lot of learning from failures, to appreciate what it can and cannot do. Hence, its participants have to understand the action–reflection–action dynamics that are built into it. Participatory development communication, as a catalyst for change, should also be accompanied by the inputs necessary for development, such as credit or capital, roads, water and technical assistance. It can establish the link between and among the development players in the community. Information and consensus are necessary conditions; but they are not sufficient to bring about broad development as desired by the community.

Likewise, devolution does not necessarily ensure effective local forest management. Policy and political and technical support have to be provided. But the use of PDC as an ingredient in the process makes it more workable, socially acceptable and satisfying in meeting the democratization and efficiency agenda of natural resource management. Moreover, PDC enables the evolution of 'participation-as-engagement' process, veering away from the usual 'participation-as-involvement' process. The latter simply implies that the 'subject' or actor is merely a participant in the process, in a position of powerlessness. The former, instead, transforms the actor into an 'active subject' and not just an object or a client. 'Participation as engagement' is a necessary condition for genuine empowerment and for creating space in local forest management, which can be best achieved through participatory development communication. The impact on the community, especially in terms of managing their own forest, would have been different had other forms of top-down communication been employed.

FROM INFORMATION TO COMMUNICATION

In Burkina Faso, as everywhere in Africa, radio exerts an undeniable attraction. In countries where the government has given up its monopoly of the airwaves, nearly all the available frequencies are now occupied by radio stations run by community, commercial or religious broadcasters. Not only do people everywhere listen to the radio, but it has taken root throughout the land, even in the most remote villages. The success of this tool, which some call 'Africa's internet', allows people to express their concerns in their own languages. Yet, this communication medium remains largely an unexploited resource when it comes to participatory development communication (PDC). An experiment conducted by the Journalists in Africa for Development (JADE) network at three places in Burkina Faso and in Mali demonstrates both the usefulness of such an approach and the importance of taking certain precautions to ensure its success.

Somewhere in Burkina Faso a farmer sits down in a studio and belts out a few songs dedicated to his friends and relatives. The man is about 50 years old. He is happy to be able to make his voice heard in his home village. This is the first time that he has ever entertained a radio audience. Earlier, he had sent his sons with a present of some yams and a chicken for the station technicians.

Elsewhere, members of a rural listeners' club have just delivered their weekly programme sheet to the station. The form is filled out in the national language and shows the broadcasts that they have heard, together with listeners' observations and their requests. The station will take these into account in its coming broadcasts.

Examples of this kind represent amazing progress compared with the way things were once done, from the time the first radio stations were established in Africa until the 1990s, when many countries in West Africa began to deregulate the airwaves. Yet, despite this freedom of speech and the openness that radio maintains with its audience, considerable effort is still required if radio stations are to become true tools for participatory communication. It is all very well to free up the airwaves or to run radio games in the villages; to greet people throughout the day by their own names; to produce and broadcast programmes about farming, livestock and the environment; and to take account of listeners' expectations. All this allows better use to be made of radio; but it is not the same thing as making radio a tool for PDC, and still less for the management of natural resources. There is no magic recipe for making radio a participatory communication tool. But the experiment that JADE conducted during 2000 to 2002 in three places (two of them in Burkina Faso and one in Mali) provides some food for thought. This chapter looks, in particular, at the most successful effort, launched in January 2001 at Ziniaré in central Burkina Faso.

Central Burkina Faso is a region of poor soils and one that faces severe water shortages. The 2001 project involved using participatory communication to share information among communities, development agents and radio producers in order to improve the management of natural resources. Besides JADE, the group includes technical experts from the agriculture and environment ministries, farmers organized as communication relay points in their communities (also called local communicators) and radio producers. The partner radio station belongs to a farmers' organization, Wend Yam.

TRADITIONAL CHANNELS: LOCAL COMMUNICATORS

The project on rural communication and sustainable development made some innovations in natural resource management by encouraging reliance on local skills, particularly local communicators. The local communicator was the cornerstone of the project and served as the interface between the relay points (*i.e.* the project representative in the area) and the community. The local communicator's many functions make him an indispensable player in any action-research effort.

In fact, local communicators perform many tasks: organizing and hosting chat sessions and supporting the research team in identifying natural resource management issues with the community. They also gather local knowledge to be dealt with by the broadcast, identify the resource persons (model farmers), organize recording sessions and gather feedback.

THE IMPORTANCE OF FOLLOW-UP

The project's impacts are due, in large part, to follow-up by stakeholders in the field. Development technicians have taken it upon themselves to share experiences, experiments and lessons learned as a way of working with the communities. They can now foster synergy among themselves and, in particular, with the radio producers. As to the radio producers, they can regularly be found collecting information in the field. They also work with the community in choosing issues, in handling them and in gathering feedback. For this purpose, they have retained the devices that were introduced (*e.g.* listeners' clubs and local communicators). Today, radio is using the participatory communication approach to go into the field, identify problems and seek solutions with local people. Moreover, thanks to local communicators who have their roots in the community, the issues of concern to local people can be inventoried and, in this way, can form the basis for radio programmes. Local people are now taking a much more active role in producing broadcasts.

In Sikasso, radio producers who were not involved in the project were, nonetheless, impressed by its achievements to the point where they wanted to use the approach in their own work. Within government technical services, as well, several managers recognized the usefulness of radio in dealing with bushfire prevention, sharing knowledge about production techniques, settling disputes over environmental management, preventing disease and reinforcing producers' groups.

Today, Radio Yam Vénégré has more direct and more regular contact with local people and is facilitating negotiation between them and the development agencies. With its experience in PDC, it is now playing an advisory role to development institutions. Moreover, it has helped a project that was designed to strengthen farmers' organizations to clearly explain the objectives of its surveys before beginning fieldwork. Programmes have been produced and broadcast for this purpose, and development officials are now working together. Even outside the project, broadcasts are being produced with other agencies attracted by the approach, something that also earns new revenues for the radio station.

By way of conclusion, we can take satisfaction from having used radio in a new way and we can congratulate ourselves on the resulting achievements. But the research team could have gone even further. It should have systematically produced audio cassettes in order to preserve the chain of communication, to reach those who missed the broadcasts and to allow certain points of debate to be revisited.

Posters or booklets would also have helped to reinforce viewpoints. With respect to institutional information – that is, community questions about the project itself and the responses of the project coordinators – use of the radio could have cleared up a number of misunderstandings. This experiment provided a good opportunity for assessing the most efficient and effective types of programming. In addition, PDC could have been compared with the conventional communication format, which is simply based on broadcasting information. Recognizing that the broadcasts occurred both before and after the group discussions, a summary of the previous broadcast could have been replayed before the discussion sessions were held. The radio station, in fact, had the right equipment, including powerful loudspeakers, to perform this task properly.

Despite the importance of rural radio, it should not take the place of chat sessions. Radio should be seen as a supplement, one that relies on the results of these sessions for input. Chat sessions must also meet certain conditions if they are to be truly participatory. In the absence of all these conditions, people will, of course, continue to listen to the radio; but what they hear will be the voices of the producers, the dominant voices of the day – it will not be a tool at the service of the community that is truly participatory.

NATURAL RESOURCE MANAGEMENT

Radio is usually seen as the ideal medium to reach people in remote areas who do not have access to other sources of information. But when committed to community participation and people's development aspirations, radio can be much more than a source of information. It can also be a powerful tool to facilitate consensus-building and decision-making at the community level, thus becoming a catalyst for change. This is precisely what Radio Ada set out to do when it decided to accompany the efforts of the people of Obane in the eastern part of Ghana. After four decades of watching their environment deteriorate, the inhabitants of that community rolled up their sleeves to restore the waterway, which used to be at the heart of their lifestyle, and regain their lost prosperity. Day in, day out, Radio Ada was by their side, voicing their concerns and acting as a morale booster. In the hands of the community, the radio also became a tool for advocacy and mobilization. The results are quite astonishing.

A Catalyst for Collective Action

Many towns and villages seem to have forgotten the reasons that they have come to be where they are. The people of Obane have not.

Obane is a rural community in Big Ada in the Dangme East district of Ghana, about 100km from the capital city of Accra. It is a brisk hour-long walk away on a hot, dusty road from Radio Ada, the community radio station of the Dangme-speaking people.

The people of Obane still remember that the waterway is the reason that their forefathers chose to live there. The waterway, the Luhue River, is a tributary of the mighty Volta River. It supplied fish, provided water for irrigation and served as a bustling transport course. The water was so abundant that Obane used to be the food basket for Big Ada. The women were, among other occupations, fishmongers, farmers, mat weavers and petty traders, while the men fished, farmed or hunted. That was over 40 years ago.

The creation of the Volta Dam during the early 1960s reduced the flooding of the fields at Obane. Weeds, trees and debris choked the waterway. The surrounding lands became barren, leaving a fetish grove with isolated trees and some patches of green to the south. Not even the shadowy line of emaciated trees tracing the meander of the Luhue River in the background is thick enough to break the line of sight to the west.

Over time, Obane became one of the poorest communities in the district. For this reason, Radio Ada has always taken a special interest in Obane. Radio Ada is Ghana's first community radio station. On the air since February 1998, the station broadcasts 17 hours daily exclusively in Dangme, the language of its listening community.[1] It is staffed by volunteers drawn from the community and trained in its home-grown workshops. Radio Ada's identity is rooted in the culture of its listening community and inspired by their desire to improve their economic way of life while maintaining close community ties. The station actively pursues a participatory development philosophy. One of the main ways in which it tries to operationalize this philosophy is through 'narrowcast' programmes. These regular weekly programmes are recorded in different communities, with the main occupational groups in the listening community – fishmongers, fishermen, women farmers, men farmers and so on. The programmes are driven by the participating community members. They determine the content and it is their voices that predominate. Programmes are presented as a continuing dialogue, where producers simply act as facilitators. The station's holistic approach to community-initiated development as 'the voice of the voiceless' in content and the production process generates trust.

Thus, the programmes serve to affirm the knowledge and experience of group members, who share what they know, what they feel and what they believe. They fuel consultation and interaction between themselves and their counterparts, as well as between other listeners in the Dangme-speaking communities. They also serve to deepen the relationship with Radio Ada. As resources permit, the stations follow up the programmes with more extensive community consultation. It was after a number of 'narrowcast' programmes had been produced with the people of Obane that one such consultation was initiated by Radio Ada. The consultation was held a few years ago, when the community was confronted with a decision to migrate or to renew their waterway.

The consultations generally involve the use of participatory rural appraisal (PRA) tools, such as the ranking-and-scoring matrix for needs or priorities. Despite the inherent friendliness of the tools, often when one asks communities about their priorities, the answer goes something like this: 'We need pipe-borne water, roads to transport our produce, electricity, employment, etc.' The litany never seems to end and tends to read like a shopping list.

In the case of Obane, possibly because of the ease and trust that had been built in their relationship with Radio Ada and almost certainly because they were at a decisive point in the life of their community, the responses were more detailed, reflective and textured. In every case, respondents had a story to tell connecting the present to the past.

The fishermen lamented that they had become goat and sheep rearers; one owns a cattle kraal. They were worried that the different fishing skills they had learned from their parents and friends were no longer of value to them and their children. The river did not exist during most of the year. It was choked, and even during the short rainy season it provided only limited access. The vast marshy lands that had boasted of crabs and reeds for weaving had all run dry. The women could not bear sending their children off to school in Big Ada and beyond without enough food to support their stay. Formerly a little rise in the tide caused natural flooding of the fields for fresh crops even during the dry season. Although Obane lacks many basic amenities, the women identified 'unity' as their most important priority for development. They were sure that they, working with their husbands and joining other neighbouring communities, could revive the economic life of the community and restore it to its former status as a food basket.

During the plenary discussion, Divisional Chief Nene Okumo was invited to share his reflections. He narrated how during the past their fathers occasionally cleared the waterway to support ecological processes and to sustain flora and fauna. Then and there, the plenary decided to dredge the Luhue River using communal labour. Immediately, Madame Adjoyo Djangma, a mother of seven and redundant fishmonger, intoned a jubilant, melodious traditional chorus. The response was infectious as others took up the chant. There were contour lines on some foreheads – more than 40 years of vegetable growth and silt to clear! The work extended over 10km, and would entail varying degrees of difficulty. In the meanwhile, the other neighbouring communities had their own priorities and there was the need for tools and other logistics. How could they possibly mobilize the resources needed?

'How can your community radio station, Radio Ada, help?' There was a measure of relief on some faces as the role of communication was put forward: 'Announce what we have decided to do... announce the day and time for the work... when others hear, they will join us... announce the names of those who have reported for work... put our needs on the radio... tell other people about the by-laws we shall make to prevent the river from choking.' The ideas kept coming. Having facilitated consensus- and decision-making on the ground, Radio Ada fuelled the communication bonds within the Obane community and connected them to other communities and institutions to support the communal labour needs. Trusting the station to honour its offer of support, the groups actively turned it into a tool for advocacy and mobilization. They announced at their convenience and free of charge their work plans – whose turn it was to work – and raised issues on air about the environment, their occupations, their lives and the project's progress.

Both the women and the men of Obane conducted the clearing, and 60,000 seedlings of red and white mangrove have since been planted. Before long, four groups from different communities – Obane, Gorm, Togbloku and Tekperkope – were working together to dredge the Luhue River. Other groups from other communities joined them to show solidarity and were duly acknowledged on air. They included communities from Dogo, Dorngwam and Atortorkope, as well as from Aminapa, Wasakuse and Midie.

District Wildlife Officer Dickson Yaw Agyeman stated: 'The radio broadcasts were a morale booster and a challenge. The Wildlife Department gave them Wellington boots, nylon ropes, cutlasses, hoses for weeding and food for work – kenkey and fish and pepper.' Their voices and active participation on air also won them other collaborators, allies and supporters. Organizations and institutes such as the Dangme East District Assembly, the Canadian High Commission, Green Earth and the Kudzragbe Clan of Ada Elders contributed to the initiative. The men's group received a loan facility of 14 million Ghanaian cedis to boost agriculture, while the women obtained a loan for 15 million cedis for food processing and marketing. In addition, a 12-seater KVIP toilet was built. Funding was through the Wetlands Management and Ramsar Sites Project, which is supported by the World Bank. The magnitude of work accomplished over these four years might intimidate a stranger; but knowing what the river and the surrounding fields mean to their lives, the people of Obane persisted. With the support of other communities and of the Wildlife Department, as well as their community radio station, Radio Ada, they now have what matters most to them.

Day in, day out, as needed, Radio Ada followed the efforts of Obane on the air. Throughout the process, it took its cues from the leadership of the community, who were constantly at the station making requests for specific programmes and announcements. Radio Ada not only helped to mobilize concrete material support for the back-breaking work of the people of Obane, but also inspired the rest of the community with their enthusiasm, perseverance and sense of unified purpose. By consistently projecting a community endeavour into the public domain, as well as through the soft cheerleading style of the radio broadcasts, even disputes over land boundaries, customs and leadership were avoided. No one could afford to risk name or reputation by swimming against what was now literally a growing tide. Today, one can travel on a boat from Big Ada through Luhuese to Obane and beyond. Year-round small-scale farming is possible by using irrigation methods, and the people now have access to freshwater. A replication is also in the offing at Totimekope near Ada Foah, the administrative centre that is the twin town of Big Ada. There, the Futue River is to be cleared in order to restore river transport, fishing and farming. The satisfaction of Radio Ada is in fulfilling its mission and sharing in the joy of the community's successes. Occasionally, however, there are more concrete rewards – perks, so to speak.

One sunny day, one of the dredging groups, led by Alfred Osifo-Doe (a professional mason), burst straight from the dredging site, sweating and jubilant, into Radio Ada, excitedly bearing a gift for the station – proof of the success of the common endeavour, they cried, fished out of the increasingly swelling waters of the Luhue. Deeply touched, but mindful of its conservation role (as well as of the practical difficulties involved), the volunteers of Radio Ada managed to persuade the dredging team to return the gift to its natural habitat. So, be careful the next time you cross the Luhue River because by now it must have grown – the baby crocodile, that is!

SUSTAINABLE FOREST MANAGEMENT INVOLVES POLITICAL AND SOCIAL PROCESSES

The past decade may come to be known by foresters as the period when they, and just about everybody else, sought to define or to prescribe sustainable forest management (SFM). Many of these efforts have painted very detailed pictures of what a well-managed forest should look like. Others have established general principles. Whatever their form, the key ingredients are sustaining multiple values through forest management, and the rights of multiple actors.

The earliest initiatives to define SFM were unilateral, undertaken by industry associations or by environmental non-governmental organizations (NGOs), and consequently were mistrusted by other stakeholders. Most of the currently accepted SFM initiatives result from multistake-holder processes, and consequently reflect a range of these stakeholders' values. They generally comprise sets of principles, criteria and indicators that have to be interpreted in detail at the local level, offering further scope for incorporating local values. They include global/intergovernmental processes (notably the UN Forest Principles 1992), regional processes (*e.g.* the Ministerial Conference on the Protection of Forests in Europe 1993) and national standards (*e.g.* the UK Forest Standards 1998). Others are led by civil society (*e.g.* the Forest Stewardship Council's Principles and Criteria). The International Institute for Environment and Development (IIED 1996) analysed 17 such initiatives and found that all had the following in common:

- sustaining yields of all socially valued goods and services from forests;
- ensuring positive impacts of forest use on different social groups;
- maintaining the state of the forest to enable continued production for future generations. These initiatives to define SFM all state or imply two premises.
  - The purpose of forestry is to provide the highly varied needs of society; this requires that demands are signalled effectively in policy and markets, and necessitates means for making trade-offs.
  - The basic principles of SFM need to be interpreted at the most local level at which specific goods and services are needed, *e.g.* the community level for recreation and firewood, the national level for major watersheds, and the global level for climate regulation. 'Sustain-ability' can be likened to liberty or justice, a goal we all understand and aspire to, but which has to be negotiated and defined locally to understand and achieve it in practice.

This all points to an emphasis on determining local values, on participation among interest groups, on using local decision-making processes, and on ensuring that forest management systems, markets and policies reflect social values. Only recently, however, have foresters been expected to focus on social issues. There is little prior experience of foresters taking on roles of social development, although there are historical examples: the famous *taungya* system developed in Burma in the 1850s was a response to local people's needs for farmland. Yet foresters are being held increasingly accountable for social conditions (in various new regulations, voluntary codes and certification). There are many unresolved arguments. To what extent can forestry be an instrument of social development? For which communities should foresters be held accountable for social conditions, those close to the forest or also those further away? Clearly, forest managers have at least to understand which groups are most dependent upon the forest and who could contribute most to the success or failure of forest management.

FORESTS PROVIDE MULTIPLE SOCIAL VALUES

The social values of forests differ widely, depending on the culture and social group in question and the roles which forests play in their livelihoods and quality of life. These values are not static, but change over time. Key change factors appear to be the following.

Shortages of, and opportunities to develop, the five main forms of capital (natural, physical, financial, human/individual and social/community capital). Where natural capital such as forest is abundant, it tends to be valued as a resource for conversion to generate other forms of less-abundant capital, and vice versa.

- In a similar vein, the absolute and relative scarcity of particular forest resources, such as timber or biodiversity.
- The availability and price of substitutes for forest goods and services.
- Scientific discovery, education and technology changes.
- Access rights, and resources such as labour that alter capacities to exploit forests.
- An individual's allegiance to specific groups and their shared values.
- Factors that change individual allegiances: campaigning and the political influence of certain groups, tastes and associated communications and media. The prevalence of individual consumerism has changed values away from those of the community.
- Changes in political culture.
- Changes in national development conditions.
- Changes in individual income.

Whilst the table begins with basic livelihood values such as food, and finishes with recreational and aesthetic values, it would be a mistake to assume that there is a universal, rigid hierarchy of needs. For example, for some groups, spiritual values are fundamental, while for others these are much less significant. It is therefore important to identify the relevant values from the point of view of the group in question. Indeed, certain groups define themselves (in part) by the values they hold in common. This is a source of many recent clashes between forest 'stakeholders'. These clashes can be exacerbated when some groups promote groundswells of broader opinion in favour of the values they hold most strongly, as we have witnessed with many environmental campaigning groups or forest peoples' groups. In other words, group values can be influenced by those of others. The outcome is, in large part, a function of power relations, something the forest manager may be able to do little about but must be aware of.

It is also crucial to be aware of the broader trends that affect social values. Policy, institutional and market conditions frequently make the realization of some values difficult. Furthermore, the role of forests in providing such values can be substituted by other sources (*e.g.* imported foods, non-forest employment and entertainment). The possibilities of substitution tend to be much greater for richer communities than for poorer, forest-based groups. Hence the need for an emphasis on working with the poor.

In developing countries, subsistence domestic products such as fuelwood, poles, fencing and fodder are a priority for forestry, the other priority area usually being food security from forest management or conversion: 'Over 1 billion people (about 20% of the world's population) depend either wholly or to a significant extent on forests, woodland or farm trees for their subsistence needs and/or livelihood'. As Filer and Sekhran (1998) point out, many people who are perceived as rather passive 'indigenous forest dwellers' think of themselves more as forest gardeners and forest developers, with food a number one concern. Forest managers ignore these pressing social needs at their peril.

Some demands for forest values are now far stronger than they were even a few years ago. For example, 'biodiversity' is a word that was coined only in the late 1980s, and the idea of carbon sequestration as a prime value is only beginning to be accepted. Yet now there are all sorts of international environmental agreements, especially the Biodiversity Convention and Climate Change Convention, with significant financial transfer mechanisms emerging to match the political obligations to produce biodiversity and carbon storage values. At present, those who are pressing for such 'global' forest values tend to be the politically more powerful countries and groups such as international environmental NGOs; much remains to be done to ensure that the production of these values also benefits local groups, rather than squeezing out their (more parochial) concerns.

It is quite a challenge for the forest manager to make decisions about the relative weights of the values held by different groups, from local to global, and how to integrate them or how to make choices between them where integration proves impossible. Three trends are helping here. Firstly, there is the trend for certain markets to recognize some social/environmental values of forest management that were previously unrewarded, for example the market for certified forest products and the emerging market for carbon storage. Secondly, there is development of more inclusive systems of forest policy and management decision-making, such as Round Tables, forest fora, and stakeholder liaison groups, where consultation with different groups about their values is on the increase (even if there is as yet insignificant participation in decision-making and management). Thirdly, there is the work by economists to categorize values and to develop methodologies to assess them on a single (financial) scale. While these methodologies remain contentious (it is not so much the financial magnitude that counts as who bears the cost and who gains the benefits) and indeed have so far made little difference to major policy decisions, especially those regarding private forests, the economic categorization of multiple values is useful to information science if not yet to decision science.

Broadly speaking, it includes the following.

- Direct-use values: where the value is derived directly from the forest, either in a consumptive manner (timber, nuts, fodder, game, fish, etc.) or a non-consumptive manner (tourism, recreation, etc.).
- Indirect-use values: where the value derives from environmental services such as watershed and social protection, carbon sequestration or biodiversity protection, rather than from the forest directly.
- Passive-use values: where value is accorded by the mere fact that the forest exists (existence value), or for the future possibilities that the forest represents (option value) or for the ability to pass the forest on to future generations (bequest values).

THE CHALLENGE OF ACHIEVING SECURITY OF FOREST VALUES

Despite the vast range of values that people seek in forests, few values have been evident in policies, markets and institutions until recently. This is largely because of the prevalence of governments and corporations in power structures and their relatively narrow requirements from forests (notably timber and foreign exchange).

Over time, the procedures of forest authorities that serve these interests have often become increasingly anomalous in the face of needs for multiple values. Their procedures have tended to become objectives in themselves, officers' rewards being based on observing hierarchical norms of behaviour and not on innovation or results, with an emphasis on narrow technical issues leading to a legitimation of increasing isolation from other forest interest groups and their values. The continued allocation and management of forests for timber alone, bolstered by such 'fortress forestry' authorities, hinders stakeholders from using forests to produce other goods and services.

There are usually several overlapping customary and legal interests attached to a piece of forest territory. Rarely is more than one set of interests, usually timber or mining, taken into consideration in forest policy, planning and land allocation. This is often exacerbated by local stakeholder rights having been removed through legislation. Moreover, because many governments tend to be weak, powerful (private sector) operators can appropriate these same rights. The net result is inequity in bearing the costs of forest use, and in enjoying the benefits. Means for dialogue, negotiation and partnerships between stakeholders that might improve the situation are weak; thus either the *status quo* is perpetuated or clashes over forest values escalate.

Consequently, many groups are experiencing insecurities in supplies of forest goods and services. These insecurities will worsen as population grows, or more particularly as the consumption of richer groups increases, whilst forest resources continue to diminish. By the middle of this century, if present trends continue, there will be twice as many people relying on a much smaller area of forest. It is imperative that demands for wood production in particular are consistent with supplying the other forest values that people need, especially those that currently have no market or can never operate within a market, such as local livelihood needs.

In agriculture, the concept of 'food security' at household, national and global levels helped to galvanize action against food shortages and famine.

The concept of 'security of forest goods and services' may help in making more precise decisions about forests, *i.e.* what goods and services forests should produce and who should share the costs and benefits. The value of the recent initiatives towards SFM principles, criteria and indicators is that they force different groups to be clear about what they seek, in terms of goods and services and the state of forest values.

The challenge, then, must be to achieve security of those forest goods and services that are socially valued at specific levels. There are significant policy and fiscal challenges regarding security of the (indirect-use) values that are required at national level, notably the production of watershed, biodiversity and other public benefits from private lands or management companies. In addition, divisions between countries of the North and South have hindered attempts to secure global forest values in the face of the assertions of many countries that forests are sovereign territory.

It is suggested that the local level, especially the community and household in forest-dependent areas, should be accorded priority, for forests are key for livelihood security and poverty reduction; furthermore, local forest producers are often best placed to generate the values that are required at national and global levels (although they will require incentives).

Forestry that improves social values and reduces poverty will be of a type that empowers poorer groups as forest stewards, improving their security of access to forest goods and services necessary for livelihoods. It will also mitigate or restrain the power of those in whose hands forests are currently over-concentrated and who are wasting or asset-stripping them.

PEOPLE'S MEANS TO ACHIEVE SECURITY OF FOREST VALUES

People have always planted and managed trees *where* they want, for the goods and services they want, *if* they have the means at their disposal, *i.e.* the technology and resources, and recognized rights, responsibilities, rewards and relationships with other groups.

Traditional knowledge

There is much discussion about the need to reconcile traditional indigenous knowledge with 'scientific' forestry in order to improve local action to secure forest values. It is ironic that, at a time when we are seeking means to sustain multiple values, many traditional management systems that do precisely this are on the decline. For example, Lacandones forest farmers in Mexico cultivate 75 species in a single hectare, and a farmer will need to clear no more than 10 ha of rain forest in his entire farming career. In Sweden, until the nineteenth century, rotational shifting cultivation sustained grain, root crops and livestock as well as timber, and the government saw it as a legitimate form of forest management. As Ghillean Prance has said, traditional forest peoples 'who depend upon so many of the species in the forest will hardly want to destroy more than the absolute minimum necessary for their agriculture. The settler from outside the region, on the other hand, is perplexed by such diversity and tends to destroy rather than use forest [diversity]'

Traditional forest management systems comprise both knowledge and rules, some of which may be transferable to new SFM approaches. The challenge is partly a democratic one: to admit local groups into policy and management decisions and to ensure that their rights can be exercised. But it is also a challenge to the prevailing systems of knowledge, which by and large are scientific and commercial, to admit such knowledge into the pantheon of accepted practices that guides our forest management.

Rights

Different groups have particular rights associated with obtaining forest values. Some rights directly concern the forest and its use:

- Territorial rights;
- Ownership of trees and other resources, such as minerals, grazing and wildlife;
- Rights of access, use and control (for specific purposes and times);
- Intellectual property rights.

Several problems are commonly associated with rights. One is that different rights overlap on one territory. Furthermore, many rights may not be established in legal documents, especially customary rights.

Other relevant stakeholder rights concern communities and their cohesion:

- Rights to protection of cultural heritage, landscapes and folklore;
- Collective rights, *e.g.* a community's rights to represent itself through its own institutions;
- Rights to religious freedom;
- Rights to self-determination and to development;
- Rights to privacy.

National laws tend to cover specific social issues linked to forests, notably rights to public consultation (planning laws), rights of access (public rights of way) and forest worker rights, etc. Internationally, there is a growing body of relevant legally binding agreements, to which countries may be party, and which eventually find their way into national legislation. These include:

- UN Covenant on Ecónomic, Social and Cultural Rights (1966);
- UN Covenant on Civil and Political Rights (1966);
- UN Declaration on the Human Right to Development (1986);
- International Labour Organization Convention 169 concerning indigenous and tribal peoples in independent countries (1989);
- UN Convention on Biological Diversity (1992). Many new laws are based on recent legal principles deriving from international consensus on sustainable development (such as the 'intergener-ational equity', 'polluter pays' and 'precautionary' principles). Laws on land, environment and forestry are being reviewed in many countries, especially regarding the links or conflicts between formal and customary rights.

Some groups are forceful in exerting their rights and/or have skills and resources to do so. Others are less well equipped. Irrespective of the stakeholders' powers and resources, forest managers are likely to experience problems if they ignore or violate these rights, and indeed claims. An informed approach is better. This would include anticipating future legal requirements by making provision for best practice in dealing with local groups, and continuous improvement of practice.

IDENTIFYING PROBLEMS IN NATURAL RESOURCE MANAGEMENT

'What do you consider to be the priority problem with water, wood, fields and the land, and how could you help to resolve it?' This was the first question put to debate in the four discussion groups, and then in a plenary session. The idea was to highlight the community's capacity to find its own means for resolving its problems before turning to other sources for assistance.

At Nagreongo, an arid region of central Burkina Faso, three issues emerged from the discussions: water shortages, soil degradation and lack of wood. A poll was taken, and water shortages received the most votes. This issue was therefore ranked first in terms of priorities. Among the reasons for the water shortage, people cited the inadequate wells, the great depth of the water table, the lack of erosion control and the absence of reservoirs. A more thorough analysis also revealed that the water was unhealthy for human consumption, and that simple techniques would have to be found to make it drinkable.

In terms of consequences and impacts, it was evident that the lack of clean drinking water led to public health problems, such as Guinea worm disease, childhood diarrhoea and other waterborne illnesses. Moreover, the shortage of water was destroying village communities as their able-bodied workforce gradually left, and was also harming livestock activities.

PROGRAMMING PRODUCED BY RADIO PERSONNEL AND LOCAL COMMUNICATORS

In a participatory approach, producing a magazine-type programme involves not only radio producers, but also local communicators. These are farmers who are responsible for identifying village interests in advance, preparing field trips, taking part in producing a broadcast and gathering feedback from listeners. In this way, a series of 45-minute radio magazine programmes was produced, essentially based on ideas from members of the various discussion groups and interspersed with a few musical interludes. This was 'raw feed' that served to highlight participants' viewpoints.

Broadcasting the Programmes

Public announcements were made, usually a week in advance, alerting people of the coming broadcast. This not only made it possible for people to tune in individually, but also, in some cases, to form discussion groups.

Gathering Feedback

The local communicators were responsible for sounding out listeners' views on the broadcasts, with the assistance of the listeners' clubs (in Ziniaré and Sikasso). Typically, a local communicator would visit the main regional town on market days, and people would pass on their opinions to him or her. In some cases, listeners with literacy skills would send their viewpoints direct to the station. Generally, the station attempts to respond directly to listeners whenever possible. It also occasionally calls upon resource persons or institutions to deal with listeners' concerns.

Lessons from the Experiment

Preliminary steps are needed to establish the sense of trust that is essential in getting the process up and running. Radio stations typically approach the communities with a preconceived idea of a broadcast. People rarely have the chance to make their expectations known in advance. Even today in rural areas, radio stations, despite their proliferation, are still inaccessible to most of the population. The fact is that when people are able to express themselves over the airwaves, this gives them some power, not only within their village but beyond. A good example of this arose in Nagreongo. Radio producers went to the village for a working session with resource personnel in order to establish a basis for collaboration and they found, to their surprise, that the entire village came out to meet them. Yet, this is a common practice, something communities do to stay in the good graces of development agencies. How could the situation be handled? It was here that local knowledge helped to avoid embarrassment. The working session originally scheduled was set aside, and the programme producers set about recording a broadcast on the history of the village, to the great satisfaction of everyone. They then held an interview with the local healer, an influential personality in the village and a great master of ceremony. These broadcasts were carried over RadioYam Vénégré in the evening.

In this particular case, the radio team turned a dicey situation into a triumph by producing a radio broadcast on the spot. This ability to react quickly to situations allowed them to win the villagers' trust, and then to hold discussions with separate groups of men, women, youth and the elderly.

Collaboration with traditional authorities, technical experts and communities (in other words, teamwork) is an essential factor for success

The participatory approach demands real collaboration among all stakeholders – traditional authorities, research teams and communities – in order to minimize the risk that things will grind to a halt if certain parties are overlooked. Teamwork produces results that reflect the great diversity of the group. In Dori, as in Sikasso, field reporting was considerably enriched by the representatives of farmers' organizations and by rural development experts. The programme producers confined themselves to formulating pertinent questions and ensuring that proper use was made of the recording equipment. Yet, this approach requires a thorough knowledge of the local setting.

RADIO GIVES ADDED VALUE TO LOCAL FOLK KNOWLEDGE

The importance of local folk knowledge has to do with the fact that it is slowly disappearing in the areas under study, without ever having been fully assessed and appreciated. Folk knowledge is not a simple store of information; rather, it has to do with the realm of understanding and of what is sacred. This is why it is important that communities are involved and that the appropriate approach, involving the following steps, is taken:

1. Pinpoint and prioritize the natural resource management problem.
2. Identify the resource persons who have local know-how relating to that problem.

3. Interview those resource persons.

The information collected must deal with the nature of local knowledge, where it was learned, its description, the way in which its holder uses it in practice, the results obtained and the constraints observed. It is also useful to develop in advance a line of argument to support any information-gathering process in order to overcome the possible reluctance of local knowledge holders and to appreciate their knowledge, which is generally brushed aside by development officers.

That line of argument might point to the fact that as the processors of this folk knowledge move on in years, it becomes ever more urgent to record and preserve what they know and to make it available to the entire population. It may also be useful to point out the paradox inherent in looking to outside knowledge, which is often costly and inappropriate, when there are readymade home-grown solutions at hand. Local knowledge, indeed, must be treated as a heritage to be preserved.

How can radio help to promote traditional local knowledge? Because it is immensely attractive to the public, radio can give to any information that it broadcasts an importance that it would not otherwise have, not only in terms of building an audience, but above all in terms of winning recognition for it. To be quoted on the radio leads to increased credibility.

SITUATIONS OF FOREST MANAGEMENT FOR CARBON COMPETE

The sites which were selected for this study are all in places where historically, degradation is the main process by which forest carbon was being lost. These are zones of rather low land value, where there is no obvious competition for alternative land use such as agriculture, because of the terrain, infrastructure such as irrigation, or because of the distance from markets. In such areas, the opportunity costs are low, and a small reward for carbon stock increase or for reduced carbon emissions may represent an attractive financial opportunity. In zones close to cities or in areas of high agricultural potential, there is more likelihood of wholesale land clearance (deforestation), and the value of carbon is probably not sufficient to counteract these processes. The areas studied were all also managed by communities, rather than individual landowners or land users. This means that the unit forest size is in the range of 50 to 600 hectares, an area in which carbon stock can easily be measured and monitored by two or three people in two or three days. This offers considerable economies of scale over individual landholdings, where each individual would have to be trained, or for each of which a special arrangement would have to be made to carry out the stock assessments.

It is also the case that forest degradation is related to uncontrolled community land uses (open access behaviour) while deforestation is more likely to be the result of individual land management decisions, as for example in the Amazon frontier where individual settlers (legal and illegal) move in and clear forest for pasture or for cropping, and the West African rain forest belt where timber companies may (legally or illegally) engage in clear felling. Thus there are several reasons why community forest management is particularly well placed as regards crediting for reduced emissions from degradation.

LOCAL TRANSACTION COSTS OF MEASUREMENT

The case studies have also demonstrated the utility of handheld computers with GIS/GPS equipment which make possible accurate mapping of the forest areas and which facilitate the storage of data on carbon stock. This seemingly 'high-tech' approach was found to be very suited to the local conditions, and village people with only a few years of primary education were able to use it after only a day's training (most of them were quite experienced in using mobile phones, which, anno 2006, are common even in the most remote villages). Indeed, local people were quick to recognize the power of such a mapping system and the additional uses to which it could be put (resolution of boundary problems with neighbouring villages etc.).

Of course, maintenance of the equipment is another matter, including recharging of the computer batteries (most of the villages were off the grid), scanning into the computer a suitable basemap, and setting out the sampling plots. For these activities it is clear that an NGO or private sector organization with some technical expertise is essential. Given this, and the cost of the computers (about $500, with a similar amount for the software) it is also clear that if the carbon stock change assessments are to be made in a cost effective manner, community forests would have to be clustered into groups, with an NGO or umbrella organization with one set of equipment assisting in perhaps 20 or 30 such forests.

Financing the Carbon

If community forest management is to be employed as a means for reducing emissions from deforestation, and particularly from degradation, in developing countries, then mechanisms to support this will be required, of which financial mechanisms will play a central role. There are issues as regards both international channels for finance and local channels for finance within the countries concerned. Here these are considered from the point of view of community forest management and how such communities could be rewarded for involvement in reducing emissions from deforestation.

International Finance Mechanisms

At the international level, two distinct modes for finance of 'reduced emissions from deforestation' are under discussion, although some combination or hybrid would be possible. The first draws inspiration from the Kyoto flexible mechanisms, and might be placed under the Kyoto Protocol through amendment or in a re-negotiated agreement relating to the second commitment period, which is expected to cover the years 2012-2017. In this approach market mechanisms are central; carbon credits are issued per ton of carbon emission reduced or sequestered, and in principle payment is made ex-post on the basis of this output. The idea is that these can be used directly to meet emission reduction goals. Within this general model of finance tied to carbon credits, there could be two possible ways in which this could be organized: either at a project level, as in the current CDM, with a project specific baseline representing the 'business as usual' rate of deforestation, or at a sectoral level, as in the proposal for 'Compensated Reductions', Party voluntarily accepts a national level target as regards emissions from deforestation, and the baseline is based on national rates of deforestation in the recent past. Either way, any reduction in the observed rate of deforestation compared to the baseline would be translated into tons of carbon, which would have to be verified and certified in some way before they can be sold. The important difference between these two versions is that in the first, the actors on the ground who are responsible for the reductions are directly involved in the deal, as with any CDM; in the second, the marketing deal is with the nation state, and nation state itself would decide how to distribute incentives or payments to encourage the actors on the ground to cooperate in reducing deforestation, or in what other ways the funds generated were to be used.

The second, quite different model for international finance is more in line with a traditional ODA approach to forestry. Financial assistance could be pledged to support efforts to counter deforestation, with a view to reducing emissions but without a direct or quantitative link to the number of tons of carbon saved, and without a direct link to the reduction targets. Such an agreement might fall directly under the UNFCCC rather than the Protocol, or indeed under another international agreement relating to forestry. Funds would be made available by states to support technical assistance and training, forest monitoring and inventory work, and other development activities with a view to helping developing countries implement policies and measures to effectively counter current rates of deforestation and degradation.

There are of course advantages and disadvantages relating to each of these models, which are much under debate at the moment. Many observers feel that ODA funds for forestry have had limited effectiveness as regards reducing deforestation in the past. Moreover they fear that if this model is used, the ODA payments will remain voluntary and not be forthcoming in large enough amounts to really change the current situation. It can be argued that only if payments are linked to legal and obligatory targets (as is the case with carbon reductions under the Kyoto Protocol) will there be sufficient pressure on countries to contribute the funds that would be required, rather than just making token payments. There is also the understanding that a market system will be the most efficient in selecting the most economical carbon mitigation opportunities. On the other hand, the causes of deforestation and degradation are not simple; they result from combinations of many different factors, many of which cannot be tackled directly or individually. A holistic, developmental approach which provides opportunities for alternative livelihoods may be the best way to deal with the problem, but to relate this directly to observable reductions in emissions could be very difficult indeed, given the many drivers and causes at work, and the variations in this in different parts of the world and indeed in different parts of any country. It can be argued therefore that it makes little sense to fund reducing deforestation on the basis of simple carbon output.

Proponents of the market-based approaches take the view that the only means to stimulate real and sufficient investment by countries is to tie this to performance and to binding caps of some sort. The current reduction quotas (average of 5.2 per cent reductions over 1990 emissions) were negotiated before deforestation was considered as a CDM option, and clearly if reducing deforestation were to be admitted as a mitigation option in a Post-Kyoto regime, these caps would have to be re-negotiated, otherwise the market value of carbon would be threatened. Some have proposed that there should be a two target system, one for reductions in fossil fuel emissions, and a separate, but equally binding one, for bio-carbon emissions, including those from deforestation.

The advantage of carbon credits tied to projects is that the savings can be pinpointed easily to particular project activities and investments, as in any CDM arrangement. The major disadvantage and difficulty of including deforestation under the CDM approach is that it is very subject to leakage, through displacement of the deforestation activities to other sites, which is virtually impossible to avoid and very difficult to account for. For that reason a national approach in which the average rate of deforestation over a whole country is measured, rather than individual sites, is much to be preferred (as in, for example, the Compensated Reduction approach).

This could, where necessary and sensible, be modified to refer to particular regions within a country. It could in principle also be expanded to cover a multi-nation region (to account for cross-border leakage which is common in many places) although this would make for a more complicated international agreement as regards sharing the credits.

Finance Mechanisms at the National Level

Whichever of the two basic models is eventually selected – a market based, carbon credit system or a system of greatly increased ODA financing to the forestry sector, focusing on reducing deforestation – there remains the question of how such funding is deployed within the country itself. It is noted that many countries have had considerable difficulty in controlling rates of deforestation in the past, not least because of increasing demand for timber products globally, but also because of internal pressures and competition for the use of land. In many cases the economic rent on retaining forest is so much lower than the potential rent from other activities that it is virtually impossible to prevent such shifts. This has of course a lot to do with the fact that the 'true' value of forest (its long term, environmental, intrinsic, and global value) is not reflected in the market system which drives such clearance. Policy mechanisms that can be used to control deforestation and degradation of forest within a country fall into three general categories, in common parlance referred to as 'sticks, carrots and sermons'. 'Sticks' are punitive measures designed to discourage activities leading to loss of forest; they include fines and other punishments for those who infringe laws and regulations designed to protect it. 'Carrots' are positive incentives such as payments for environmental services, or other rewards for not destroying forest. 'Sermons' refers to a wide range of informational activities which in different countries may be referred to as 'raising awareness' or 'education' of local people about the value of forest, or 'motivating' people to reduce their forest-destructive practices, through persuasion. Naturally, forest policy can rest on a combination of carrots, sticks and sermons.

Finance is required for measures in each of these categories, as well as to monitor closely the actual situation as regards deforestation/degradation. Clearly, whatever the package or mix of measures selected, some finance will be needed centrally to pay for the overall management and for activities that need input from the centre, while other finance will need to be distributed, particularly in the case of 'carrots', but also for the implementation of measures of the 'stick' and 'sermons' sort. The appropriate balance will be different in every country.

Here it is important to consider deforestation separately from degradation, since the two processes may have quite different drivers, and thus may require quite different counter-measures. Apart from other reasons, as already noted degradation often needs to be tackled through an organization at community level, since it affects the common property resources, while deforestation may be more often associated with individual or state land holdings and would need a different organizational approach.

THE FATE OF CARBON FUNDS IN THE FUTURE

If the local community were to be rewarded in financial terms for the carbon saved as a result of their forest management, would principles of equality hold, and would the poorer and less powerful part of the population, and women, benefit at all? The preliminary findings from the case study in Baghmara Buffer Zone Community Forest as regards the current distribution of benefits indicate that particularly as regards financial benefits, it is the richer parts of the population who gain most, even though most of the poorer people (Musahar excepted), and women, get a fair share of the products in terms of fodder and firewood. This outcome is not surprising since it is the men of higher caste and income that get to make the main decisions, despite the idea that the FUGs are supposed to be run on democratic lines.

Whether this pattern would be repeated if a greater financial reward is entered into the system through sale of carbon sequestered or deforestation avoided, is not entirely clear. For example, one of the main reasons why the richer families benefit is because they are able to take loans for certain equipment from the community forestry fund; they have the means to match loans and collateral against the repayment. If money for carbon were not handled in the form of loans but (at least in part) distributed to members directly as an annual payment, then this problem should be overcome, and indeed the poor people would stand to earn a welcome, if small, additional income. It remains to be seen whether rules on membership would be tightened to limit membership in some way, if the financial rewards from carbon credits were considerable.

At present membership is all inclusive. All this implies is that if equity goals are to be taken seriously, some serious consideration needs to be made regarding how the whole system of rules and procedures for internal payment of carbon services is to be designed, and that particular attention needs to be paid to how the needs and rights of the "weaker groups" will be guaranteed.

CHALLENGES OF COMMUNITY-FORESTRY BASED CARBON PROJECTS

Small local communities worldwide are those most immediately at the front lines of development efforts and climate change impacts. Droughts, fires, floods, cyclonic storms, sea level rise, and changes in the food chain are all hazards associated with climate change impacts. Communities around the world are learning how to adapt; in the absence of such adaptation, they remain exposed to hazards that will threaten their lives and their livelihoods. But these communities – sometimes organized through formal or informal Community Based Organizations (CBOs) – have become accidental partners in the fight against climate change effects. Their partners include international NGOs, multilateral development banks, and bilateral agencies, various levels of local government, other CBOs, and the private sector in mitigating the effects of climate change. But in so doing, they often find themselves in another world.

While communities have been adapting to local weather effects since the dawn of time, the combination of degraded ecosystems, rapid social changes and climate change, often overstretch the capacity to adapt. In this context, climate change mitigation (regarding International Panel for Climate Change definition) is an opportunity to increase the adaptation capacity through related financing mechanisms and evolving partnerships especially on high scale implementations. Thus the concept is new for them, as it is for many of us.

In less than one generation, communities in some of the poorest countries in the world now find themselves engaged in a quest that the previous generation could not have imagined: the quest to sequester carbon. How do you sequester something you cannot see? Why would you do that? How do you know if you can store it forever? And why would anyone want to pay me for something that I then seem to get to keep? For those of us in the global pursuit for carbon, we tend to treat it little differently than any other commodity. But to the communities at the front lines, the reasons for the pursuit are not always obvious.

This paper describes a brief journey through a handful of projects that have taught us some lessons on the challenges of dealing with community forestry - based implementation and monitoring of carbon projects. We take our examples from mangrove and agroforestry projects across four sites on two continents. Our first challenge is communicating the concepts and methodologies of carbon markets to individuals, communities, and other stakeholders on whom we will rely for the next two decades to engage in a process that will involve defining, designing, implementing and seeing through a project that stretches local capacity to its limits.

Why will a tree planted on this side of the river count as carbon captured, but a tree planted on the other side not count? Our second challenge is discovering the incentives, motivations and needs of our community partners through participation that allows us to align and reconcile different objectives.

Trees bear fruit, decrease wind damage, and capture carbon by producing biomass, but what if there are trade-offs between forest livelihoods functions and carbon gains? Our third challenge is monitoring the performance of the process and the impacts of implementation efforts alongside community partners. We know that the communities have an interest in the outcomes, but how can methods be aligned to the realities of limited capacity, limited resources, and sometimes arbitrary criteria? Moreover, how can monitoring be completed in what is at times a difficult environment?

Our findings are organized around these three themes: process, participation, and performance. In addressing them we first provide a brief overview of the projects from which we draw our experience.. We then cover the intricacies of different methodologies and frameworks for looking at carbon projects: these include the Clean Development Mechanisms protocols and other voluntary carbon market protocols such as Verified carbon Standards (VCS) or Gold Standard. The methodologies themselves present challenges that influence process, participation, and performance. We then look at some of the key issues within each of our three themes, finally closing with lessons and selected recommendations.

PROJECT BACKGROUNDS

Our projects draw from a selection of Livelihoods Fund (LV) activities across Africa and Asia. Although they are all forestry projects, they are far from being duplicates of each other. In fact, their selection shows us immediately that the idea of community can even differ from one project to the next.

Senegal "Oceanium"

Since 2009, the Livelihoods Fund in partnership with the Senegalese NGO Oceanium have planted over 100 million mangrove trees on 10,500 hectares of land in 450 villages in Casamance and Sine Saloum in Senegal to replenish their food ecosystem, the fish and shellfish resources, and to lead the fight against salinization of agricultural land. Primarily this project has been launched because of the vision of one man – Haidar Ali – asking the population to plant mangrove trees to support their own livelihoods: "Plant a tree in your head" was the basis for community mobilization.

The role of private sector engagement – such as Livelihoods Fund – within such a project is imminent but requires an investment perspective to achieve its goals on time, in quantity, and in quality everywhere at a large scale; this implies that communities must adopt relevant practices for replicating efficient pilot models. As an example let us first describe soil fertility quality control, which is a key parameter for replication. Scientists can identify the key indicators: salinity, water flow, sand content, but they cannot generally identify the adequate tools for community people. Villagers are not accustomed to using tools such as refractometers.

That means that the suitability of the soil quality has to be measured with the villagers in other ways. We found indicators which are few, visible, tangible situations easy to appreciate by the villagers and correlated with soil quality: remaining presence of trees, colour of the soil, highest tide coverage. And then depending on this, each plot is clustered as Green (3 indicators OK means very good conditions for growing), Yellow (1-2 indicators OK means good conditions for growing), or Red (less than 1 indicator OK means Bad conditions for growing).

Another example addresses monitoring of tree stem density. How does one measure tree density on 10.000 ha? The solution is in the process of planting itself. Once again scientists have designed an optimum situation by planting initially 5000 propagules/ha. But practically how does one design a regular process to plant one seed every 2m*1m on 10.000 ha? The solution is a rope where you put a knot every 2 meters and during the planting session each villager is positioned along the rope at each knot and every step they plant one seed.

These indirect "community parameters" have to be calibrated to give confidence in the process. For this reason, each campaign is subjected to an external audit by experts, using a dual set of measurements on sampled plots.

India "NEWS"

India's Nature, Environment, and Wildlife Society (NEWS) is at the forefront of implementing a 6000 ha mangrove planting scheme in India's Sundarbans. While many people know of the Sundarbans forest as one of the largest remaining natural mangrove ecosystems in the world, what they do not realize is that almost 4 million of India's poorest population try to eke out a living behind a 150 year old system of dykes on about 70 islands. The dykes protect them from flooding, but breaches occur from poor maintenance, or from regular tropical cyclones.

The Bay of Bengal is home to the deadliest cyclones recorded in human history, and increasing sea levels coupled with global warming promise higher storm surges. Mangroves provide a life-saving function as a "bioshield" through dissipating wind and wave energy, while also supporting important livelihoods such as fisheries, honey collection, and a large variety of sustainable uses.

The NEWS's project was initiated in 2010 as an extension of already successful livelihood activities involving health care services and income generating schemes. Through mobilization of women's groups, plantations were designed to accept seedlings from nurseries or direct planting. While activities started in one part of the Sundarbans region, the project (registered under CDM) needed to go further afield to find suitable CDM eligible planting conditions and willing villages. Government authorities of India's forest department, as well as local government officials, have been instrumental in providing permissions relating to land use and carbon rights. To date, some 10 million trees have been planted focusing on 7 species such as Avicennia alba. and Rhizophora mucronata., Bruguiera gymnorrhiza.

Significant logistical challenges are posed by seasonal storms, difficult transportation conditions, and residual competing demands for land use. Concerning competing demands for land use, as an example of participation, during the validation visit in January 2012, the independent validation expert noticed that there were cows on some mudflats that were not part of NEWS's activities, but that there were no cows on the NEWS/Livelihoods plot. He asked the local village head how that was possible. He explained that there was an old woman that patrolled the area with a stick. "So," said the validator, "she hits the cows?" "No," was the reply, "the cow would just come back. She hits the owner of the cow."

India "Naandi"

The overall goal of the project developed by Naandi Foundation is to improve the livelihood of small and marginalized tribal communities in the Araku Valley of Visakhapatnam district in the state of Andhra Pradesh in India. Various horticultural tree species will be planted in a phased approach on 6,000 ha in this CDM project activity. Seven different tree species groups are distinguished based mainly on similar growth conditions (similar biomass accumulation rates) and planting densities. All groups will be planted on the 6,000 ha, with different species composition in the mixed stand models. Coffee plantings will be introduced after 3 years on 3,000 ha.

The plantations are all located on degraded land with very low plant cover (more than 57 per cent of the land is classified as barren land.). The horticultural trees will sequester carbon and will turn this degraded, low carbon lands into a fruit and herb bearing secondary forest. The transformation of a low carbon landscape into a high carbon sequestering, multiple use landscape will improve food security, nutrition and provide additional income for the community. The project area will cover 302 villages and approximately 10 planting sites per village, *i.e.* on in total 3000 discrete areas of non-forested land, each area has a geographical identification.

The planting process is developed as a "one shot process" and is a key performance innovation for the initial sequestration of carbon. It has been developed by villagers together with Naandi field officers.

First, the pits are filled with topsoil. A biodynamic site fertilizer strategy is applied during the time of planting. Cow Pat Pit (CPP) manure consists of fermented cow dung combined with eggshell powder, basalt dust, and herbal preparations, which is retained in a brick pit for a period of 3 to 4 months and mixed with water (1 kg of CPP manure in 40 liters of water per acre before application).

The participating farmers themselves add 5–6 kg of compost for each planting pit Approximately 250 g of vermin-culture compost is mixed with the topsoil while filling the pit for planting

Healthy, straight-growing grafts from reliable sources are planted at the center of pits with the ball of the earth intact during rainy season. The planting is done for balls in such a way that the roots are not expanded. Plants are watered immediately after planting. In the initial one or two years, a stake is provided to make them grow straight.

Indonesia "Yagasu"

Indonesia's Yagasu Foundation has an elephant as its symbol: the brand is well recognized in Northern Sumatra and Aceh because of the efforts of Yagasu.I Its few domesticated elephants, rescued lives and protected property in the aftermath of the December 2004 tsunami that struck the northern coast of what is now part of this project area.

In a project that nominally includes 5000 ha of mangrove plantation, about one-quarter is in or adjacent to some of the most heavily devastated shorelines in the world. Where once stood houses and buildings and rice fields, now stand a few trees, a tidal marsh, and occasional offerings in memory of the 170,000 people that perished during the tsunami. But these areas, as many others along the North Sumatra coast, were converted decades ago to fish and shrimp ponds in a national scheme intended to improve the lives of coastal people: mangroves were cleared, fishponds appeared. Coastal erosion and salination, however, undermined these schemes and most ponds were eventually abandoned. The abandoned areas are now a target of this Livelihoods carbon project.

Working through some [90] communities, Yagasu is re-establishing a "Coastal Carbon Corridor" intended to improve fishery yields and trap sediment. The Livelihoods sponsored project commenced in 2011, building on a number of pilot activities that Yagasu had already successfully completed in the area. The project has strong support from Provincial ministries responsible for environment and forestry, and registration of the project nationally under CDM would represent one of the first of its kind for this forest-rich country. By early 2012, some 1.6 million plants (including Rhizophora mucronata, R. apiculata and R. stylosa) were planted along 572 hectares of the coast.

Based on initial results Yagasu appears to be on the right track from the initial visits and meetings. After showing us their site, we were served cookies and drinks made from mangrove products. The products had already gone through various early stages of nutritional testing and were being sold in Indonesian airports to departing tourists with all proceeds going back to the local communities.

CARBON METHODOLOGIES

The concept of carbon sequestration in a forest ecosystem is relatively straightforward. A plot of land has growing trees that sequester CO_2 from the air and it eventually is stored in the trunk branches, above and below ground in the form of carbon in its biomass. The dry weight of the wood contains about 50 per cent carbon. Credits are typically issued for the carbon sequestered expressed in terms of tons of CO_2 equivalent (tCO_2e). The role of a methodology and the guiding carbon standard is to determine when a certain activity qualifies for carbon sequestration and, if it does, how and when the carbon should be counted or credited. A key idea in most methodologies is that only additional or incremental carbon should be counted, and it should be permanently sequestered. Another key concept is that the carbon credit becomes a unique tradable commodity with market value, meaning it has an identifiable source and becomes in effect a private good rather than a public one.

Depending on the carbon standard applied it follows different compliance procedures and is also named differently as stipulated below. We strongly believe that by extracting from different methodologies key requirements, carbon from most restored and dynamically developing ecosystems can be accurately accounted. Numerous formal and informal methodologies exist to address the eligibility and valuation issue, but we will characterize those treated here as drivers for soil management efficiency, sustainability (social and economic) for communities in Section IVUNFCCC CDM related carbon credits from forestry generate temporary certified emission reductions (tCER), but despite the spirit of the convention that these should be traded the EU member states did not allow their use in the European Trading System (ETS).

Therefore, the fungibility is limited but companies use them but have to replace them after 20 years or latest after 60 years when registered twice. A number of methodologies have been specifically developed for forestry activities considering not only tree biomass but also carbon stock changes in soils and below ground biomass. In the Livelihood projects this internationally accepted standard will be used to validate the project. The following drivers for communities have been met in each of the Livelihoods projects described above.

- Areas have to be non-forested since 1990, which presents the communities a chance to restore degraded and unproductive areas.
- The project duration (at least 20 years) refers to long term commitment of communities to preserve restored areas, bind communities, project developers, and local authorities. This concept is defined as permanency by UNFCCC.
- The UNFCCC procedures require the formal approval by national authorities, which can be considered as a constraint compared to voluntary carbon standards where this is not required. Nevertheless, it's a way to create a durable relationship and strengthens the legal recognition of the project among the host government, project developers, communities and national authorities. 1) Reducing Emissions from Deforestation and forest Degradation (REDD+) is an UN sanctioned system (UN-REDD a respective support programme) but currently respective carbon credits cannot be generated and traded under the UN system (only in the voluntary carbon market;. Respective schemes cover in general large areas and communities are involved to the extent that they must be engaged to stop deforestation and forest degradation. They also can benefit from a share of revenue generated by the REDD initiative. The plus of REDD+ indicates that also carbon stock enhancing measures such as biodiversity conservation and timber management are eligible. Eligibility has to be demonstrated using a structured documentation of the baseline situation including information on the land use in the past and what would be the situation without the project against which the additional carbon enhancing activities can be claimed. Communities need to understand the respective logic and its implications, *i.e.* can credits be claimed at all and which amount of

credits that can be generated. Carbon finance provided to communities for carbon credits is considered as a way to pay for environmental services. However, there is a distinct difference between services such as provision of clean water (which can be traded continuously) and carbon where each ton sequestered can be transacted only once, hence when trees are mature no more carbon credits can be generated. Reflecting the need to align incentives for the community and the climate an up-front investment that generates livelihood benefits such as fruit trees or fish habitats and provides climate benefits permanently even after the carbon crediting period expires is preferred. 2) Originally the Voluntary Carbon Standard but now the Verified Carbon Standard VCS, was developed for voluntary carbon transactions and was developed from the experiences gained from the CDM mechanism (CITE). The VCS seeks to establish high quality carbon credits with the same fungibility as non-forestry related CDM credits, but with potential to consider REDD+ projects and also agricultural soil carbon projects. The VCS is recognizing the permanency of the carbon credits from all forest and other land use projects. Carbon credits are issued and adjusted according to the risk of non-permanency of the projects through recognition of buffer amounts. These buffered carbon credits are constituted by a part of issued carbon credits of each project in a mutual way by VCS to mitigate the risk of failure of some projects. Eligibility requires a structured risk analysis to evaluate the "buffer" level involving since the starting point of the project communities on risk allowance. 3) A standard addressing the community oriented non-carbon project benefits is namely the community, biodiversity and climate adaptation (CCBA) benefits. This wider scope is of great interest for investors, which require that projects sequester carbon and generate positive direct and indirect biodiversity and community co-benefits. Gold status is awarded when the project conserves globally rare species and has exceptionally strong community and climate change adaptation benefits. Monitoring processes have to integrate related community, biodiversity and adaptation benefits, which in fact are a challenge since the benefits are diverse and difficult to quantify (*e.g.* biodiversity benefits). However, we think it is very important to quantify respective benefits in a robust, but focused manner to make the monitoring of the impacts by the communities themselves as a sustainable way to make them involve in the conservation of the ecosystem. 4) Simplified informal credits or offsets. These represent a broad class, which do not necessarily involve formal tracking. They include a wide range of initiatives: simple tree-planting campaigns funded by individual patrons not necessarily interested in the carbon; un-audited offsets by corporations or individuals; or, audited offsets through strict methods that are never intended to be traded. We do not spend much time on this category, but we note that it is the type that has the easiest intuitive connection to small communities in developing countries. In the minds of many, a tree planted or protected anywhere should receive a credit. The standards above generally require robust and therefore complex methodologies and independent third party verification to generate carbon assets.

The most noticeable differences among these methods include issues of eligibility, scale, and scope. Eligibility refers to the general compliance of a project area to a specific standard. In the CDM the definition of "forest" differs from one country to the next, within an internationally defined range, *i.e.* minimum canopy cover 10-30 per cent, potential tree height 3-5m, and area 0.1ha-1ha

The scale of a project is a concern in many methodologies. Under CDM, both large and small scale projects exist, defined by a threshold of the amount of carbon sequestered; a forestry project removing fewer than 16,000 tCO_2e a year can qualify for simplified procedures that reduce administrative and monitoring requirements. But the thresholds are quite low; all four projects in this paper are of a size that requires large-scale methods to be applied. Some methodologies allow extensions to be made, such as the CDM rules governing "Programmes of Activities (POA)" which in effect permit indefinite number projects to be replicated elsewhere in a country once the initial project has been scrutinized and validated. Scale can also come into play inside a project: the CDM methodologies for reforestation require that 67 per cent of the total project area is identified and delineated before credits can be earned: this is at times difficult for community projects requiring multi-year planting programmes.

Finally, new standards are evolving *e.g.* to better capture co-benefits (*e.g.* biodiversity and socio-economic) within project boundaries. Meeting related eligibility requirements may underline the "charismatic carbon" associated with the project and may warrant a price premium and will improve the project design or its documentation.

[B]THE GLOBAL CARBON TRADE AND INDIA'S FORESTS[/B]

Under the clean development mechanism (CDM) conceptualised in the Kyoto Protocol and put into practice to mitigate global warming, industries make billions of dollars by selling tonnes of carbon they 'conceptually' save. The concept goes something like this: businesses in the developed world 'invest' in helping businesses in the developing world set up and run 'clean technologies' that emit less carbon. The price in the international market for a tonne of carbon has been agreed on by all parties to the UNFCCC (United Nations Framework Convention on Climate Change) which monitors this new-world trade. The amount of carbon saved through clean technologies, compared to that in the business-as-usual scenario, is provided as compensation (according to the price fixed) to the 'implementing' business in the developing world by the 'investing' business in the developed world. In return, the latter is allowed to emit the same amount of carbon back home.

It's a little baffling as to how this mechanism actually reduces net global emissions, and how it addresses the impending perils of global warming! There are also huge question marks over whether the so-called CDM projects (there are about 1,200 now in India) actually emit less carbon. The day I visited villages around the Hindalco smelter near Hirakud dam in Orissa, the villagers were fuming (some were even crying) over the destruction, the previous day, of full-grown paddy sown on tens of acres of farmland due to pollution from the factory.

In a document released in August 2009 by the Ministry of Environment and Forests (MoEF) titled 'India's Forest and Tree Cover: Contribution as a Carbon Sink' (it's actually the abridged version of a technical paper of the same name produced by the Indian Council of Forestry Research and Education, and re-released by the MoEF), minister Jairam Ramesh writes: "Forestry is at the centrestage of global climate change negotiations. This is because forests have the potential to be a carbon sink as well as a source of carbon emissions… It is India's view that we need an agreement on a comprehensive framework for compensation and positive incentives for forestry as part of the ongoing climate change negotiations."

This position is perfectly in line with the Indian state's long-held belief that forests are the source of a never-ending stream of revenue. Now, in the context of climate change negotiations, India's forests have opened up an 'enormous opportunity' to maximise this revenue.

How will it work?

The technical paper by the ICFRE — which is going to be the basis for climate change negotiations involving India's forests — has all the arguments. "(Forests) absorb CO2 from the atmosphere and store carbon in wood, leaves, litter, roots, and soil by acting as 'carbon sinks'… Forests, by acting as sinks, are considered to moderate the global climate."

The MoEF document, which basically quotes the technical paper, says: "From 1995 to 2005, the carbon stocks stored in our forests and trees have increased from 6,245 MT (million tonnes) to 6,662 MT, registering an annual increment of 38 MT of carbon or 138 MT of CO2-equivalent… Our estimates show that the annual CO2 removals by India's forest and tree cover is enough to neutralise 11.25 per cent of India's total greenhouse gas emissions (CO2-equivalent) at 1994 levels… This is equivalent to offsetting 100 per cent emissions from all energy in residential and transport sectors, or 40 per cent of total emissions from the agriculture sector. Clearly, India's forest and tree cover is serving as a major mode of carbon mitigation for India and the world."

The catch comes at the end: "Putting a conservative value of USD 5 per tonne of CO2 locked in our forests, this huge sink of about 24,000 MT of CO2 is worth USD 120 billion, or ₹ 600,000 crore (6,000 billion). Incremental carbon under scenario 3 (in which India's forest cover increases at a rate higher than the historical rate, by 2015) will add a value of around USD 1.2 billion, or ₹ 6,000 crore (60 billion) every year to India's treasury of forest sink, assuming a value of USD 7 per tonne."

A National Forum of Forest People and Forest Workers position paper titled 'Imaginary Sinks: India's REDD Ambitions' states: "This opportunity becomes a windfall with the advent of REDD, or the proposal that CDM projects should cover the Reduced Emissions through Deforestation and Degradation scenario, which means if a project reduces any emissions caused by deforestation and degradation of forests, it should get CDM status. India now contends that it should be given carbon credits for both its old and new forests." Sensing an opportunity here, the country has an ambitious programme of raising 30 million hectares of new plantations on both 'degraded' lands and farmlands, unleashing a fresh burden on food security.

The NFFPFW position paper throws some light on this: "Instead of linear and wholesome 'constant' constructs, Indian forests (like any other tropical forests) are part of a larger dynamic and ever-changing socio-political and socio-ecological discourse (or a set of such discourses). Mathematical calculations and simulated models to project sequestration of carbon in forests can never be expected to assimilate the innumerable, essentially asymmetric and 'truant' variables that such discourses contain. Neither hypothetical baselines nor imaginary 'future' scenarios can explain/interpret/predict contextually related but often spatially separated sets of uncertain social, political, and ecological events influencing 'deforestation' events. This methodological impossibility, coupled with doubtful and unverified official forest cover estimates, makes an estimation of all carbon stocks in Indian forests downright impossible."

Forests are natural resources, and global warming is largely a man-made phenomenon. If we were actually serious about containing emissions in order to address global warming, only an incriminating mind would justify standing forests as a passport to emitting more. The somewhat vacuous argument is that since we have so much standing forests that absorb so much carbon, industries only have to pay a price and then happily emit as much greenhouse gas as they want! There is already huge pressure on India's forests thanks to the 'consumerist' development model we are so aggressively pursuing.

India's forests get CDM status, by accepting whatever carbon sequestration calculation ethod, who reaps the benefits? The countless forest communities that have protected and kept the forests alive? The forest department that has always considered India's forests its private fiefdom?

It is often said that if the CDM regime is really about rewarding those who have a small carbon footprint, the world's forest people and rag-pickers would have been the richest communities on earth by now! But that will never happen. Because the CDM regime is all about 'business'; it has nothing to do with emission reduction or global warming. The latter is just a 'premise' to strike profitable deals. The unprecedented sincerity with which the Indian government is pushing for India's forests to get CDM status is precisely to clear hurdles in the path of privatising forests. For, with so much money at stake, the forests cannot be left in the hands of those who cannot 'manage them efficiently' ('efficiency' being a typical capitalist construct).

And so, ironically, history is about to repeat itself, reminding us of Dietrich Brandis who came to India to 'manage' our forests scientifically some 150 years ago. Now, private players are waiting to 'manage our forests efficiently'.

The attempt to privatise India's forests is not new; the pulp and paper lobby has been trying since 1992 to lease 'degraded forests' ('degraded' according to the forest department's definition) in order to meet growing demands for raw material. The lobby has been arguing that industrial demand could, to a large extent, be met by raising 'protected' private plantations on forest lands that are degraded.

The Indian government, in fact, did try to push a Bill in 1994 facilitating the handing over of 'degraded forests' to industries. Besides stiff resistance from community groups, the move was opposed by the Planning Commission which constituted an expert committee, chaired by N C Saxena, to look into the matter. The 'Saxena Reports 1998' categorically refute industry's claims that degraded lands do not support biodiversity and are not used by communities. The reports argue that leasing out forests to industry would prove to be both ecologically and socially harmful, and would be an injustice to communities that use the forests for their livelihood and other purposes. It clearly states that no forests in the country can be labelled 'absolutely degraded'. In 2003, industry replied with the 'Re-greening India' report (published by the Confederation of Indian Industry), which made out a strong case for the lease of forests to the private sector. The CII report mentioned the possibility of earning carbon credits, as trees sequester carbon.

Then the World Bank joined in, advocating for private players to manage India's forests. 'World Bank Forest Strategy 2003' says that most communities fail to garner the full potential of forests, even though they are successfully protecting them. The case for private players comes in very strongly in the assessment that the total forest income from commercial timber, bamboo, and non-timber products in [I]improved [/I]forests is expected to rise from an estimated USD 222 million in 2004 to approximately USD 2 billion (2,000 million) by 2020. With so much money locked in our forests, it would be blasphemous to even suggest that the private sector would resist the temptation, especially at a time when everybody, from government to the middle class to the mainstream media, believes that traditional communities have no right to sit on 'profits' that are due to the private sector!

One can well imagine what turn the 'market' will take once REDD and the forestry CDM idea are put into effect. Going by an ICFRE technical paper estimate, all the stored carbon converted into credits would amount to a staggering ₹ 6,000 billion! The NFFPFW adds: "This is a new market, which, besides the Indian government, all shades of carbon brokers, consultants, and investors eye expectantly, and for good reason. Even in the non-Kyoto voluntary market, forestry credits are fetching about USD 8 per credit (the UNFCCC has not yet issued credits to forestry projects). This can increase enormously with UNFCCC approval."

THE EVOLUTION OF COMMUNITY FORESTRY IN PRACTICE

Early initiatives understandably tended to become focused on those issues perceived to be of particular importance. Of these, the fuelwood shortage became far the most important. It has been estimated that wood provides roughly 20% of all energy in Asia and Latin America, and that about 50% of all energy in Africa is wood generated. The early analyses emphasised the huge numbers of people affected, the apparent "gap" between demand for fuelwood and sustainable supplies - and a seemingly exponential growth in demand with population growth - and the growing burden placed on users of having to search ever further afield for fuelwood and of having to divert crop and animal residues needed for soil working or as livestock feed to fuel use.

A general conclusion from this earlier analytical work was that existing wood stocks were being widely mined to meet fuelwood demand, that there was no feasible large scale alternative to wood fuel - other than other biomass such as crop residues and dung - and that the principal means of averting growing shortages, and the attendant deforestation and human suffering, was to initiate widespread planting of additional trees. When applied to individual countries and regions such analyses resulted in programme targets on a formidable scale. For example, a major World Bank study for sub-Saharan Africa estimated that tree planting would have to increase fifteen-fold in order to close the projected fuelwood gap in the year 2000. As a consequence, a very large part of the initial investment in community forestry was in the form of afforestation projects to increase fuelwood supplies. Many early community forestry projects therefore took shape more as a response to an energy supply/demand problem, rather than to the problem of meeting local needs for trees and tree products.

The design of early community forestry projects was also strongly influenced by the fact that the more striking existing programmes were organised in the form of communal activities - the village woodlots in Korea, the panchayat woodlots of early social forestry programmes in India, the village afforestation programme in Tanzania. As is suggested by their titles, all of these were concerned with creating new plantations, rather than with management of existing forests.

These perceived imperatives had the effect of concentrating the early community forestry effort to just a rather narrow part of the spectrum of linkages between people and trees and tree products that had been identified initially - namely to establishment of new plantations and to fuelwood. There was little in the first generation of projects which was concerned with outputs from existing forests, or with the food, employment and income dimensions. Indeed, the concern with meeting subsistence needs for fuelwood even led, on occasion, to attempts to exclude income generation activities from the project design on the grounds that it was inconsistent with the perceived subsistence aims of community forestry.

BROADENING THE KNOWLEDGE BASE

In subsequent sections of this document we explore the reasons why these, and other, unforeseen developments occurred in the early years of community forestry. However, a number of salient lessons which began to emerge early in the period deserve to be mentioned at this point.

One is that production and use of tree products at the village level is in practice often embedded in complex resource and social systems, within which most of the factors that affect our ability to intervene with forestry solutions are of a non-forestry nature. They are primarily human factors, connected with the ways women and men organize the use of their land and other resources. They therefore require situation-specific approaches and are unlikely to be successfully tackled by generalised solutions or approaches that address only a single element of the situation.

A second is that earlier analyses of the nature of women's and men's dependence on trees and tree products was in some respects incorrect or incomplete and the solutions identified were consequently inappropriate. As is discussed in some detail later, this is particularly the case with respect to responses to declining supplies of fuelwood and to the attempts to insert interventions that conflicted with the existing social and institutional framework within a community. A third is that even projects which have sought to identify local needs, aspirations and possibilities have in practice done so more on the basis of the views of planners and others from outside than on the local people themselves. Dialogue to achieve local participation has all too often started only after the project design has been finalised and is in place. Though the concept of participation took root quickly, in practice it has been, and still is, more frequently preached than practised.

A fourth is that "community forestry" has suffered from considerable confusion and lack of clarity as to its nature and purpose. The use of this umbrella term seems on occasion to have obscured the fact that the objectives set for projects to support community forestry have varied considerably. Project design, and performance, have frequently suffered from a lack of clarity as to which of these objectives were being pursued or had priority. Although some among multiple goals may be congruent or reinforce each other, others may be in conflict. Planting trees to meet environmental objectives such as soil protection is unlikely to produce sufficient output of saleable products as to be economically attractive to the farmer. Similarly, tree growing designed to generate income is unlikely in itself to benefit those with little or no land. Production to meet both subsistence and market needs is unlikely to be achieved with a single production model. Projects originally designed to meet a production goal are unlikely to be equally successful at achieving a subsequently added social goal, such as favouring the poor, unless they are appropriately restructured. And so on.

COMMUNITY-BASED NATURAL RESOURCE MANAGEMENT ORGANIZATION

A Comité de Investigación Agrícola Local (CIAL) or Local Agricultural Research Committee is a research service belonging to, and managed by, a rural community. The research group is made up of volunteer farmers, chosen for their aptitude in experimentation. The CIAL links farmer-researchers with formal research systems, increasing local capacity to exert demand on the formal system and to access potentially useful skills, information and research products.

A CIAT team developed the CIAL concept, which was first tested in Colombia. Five CIALs were formed in Cauca, Colombia, in 1990 and NGOs began their involvment in 1991. During 1995–1996, the CIAL project spread to other countries, involving rural communities, GOs, NGOs and universities. At present, over 250 CIALs are active in eight Latin American countries.

Experience has shown that the CIAL can benefit the wider community as well as individual CIAL members. Mature CIALs often launch small businesses, selling improved seed or other products or services. Many take on a broader role in the community, seeking access to credit and training, preparing and submitting proposals and acting as ambassadors in relationships with research and development (R&D) actors.

Approach

The main CIAL principles are:

- Building on experience and learning-by-doing to generate knowledge.
- Mutual respect and accountability and shared decision-making as the foundations of the relationship between the CIAL, the community and external actors.

- Risk shared by partners in the research process.
- Systematic comparison of alternatives for improving farming and natural resource management.
- Research products are public goods.

Each CIAL has at least four elected members (many expand beyond this) and a facilitator. The facilitator may be a trained agronomist from a supportive formal research centre or university, an extension service or an NGO, or a trained farmer, who has served on a CIAL. The facilitator plays a key role in developing the CIAL's competence in the research process, and provides feedback on farmers' priorities and research results to formal research and extension services.

The CIAL process cycles through the following stages:

- Training of facilitators and site selection by committed research organizations, extension services, NGOs, or community-based organizations.
- Mobilization though an initial community meeting and regular interaction between the CIAL and the community as the research process unfolds.
- Election, when the community chooses farmer researchers who are interested in experimenting and willing to provide a service to the community.
- Diagnosis, when the topic or question for experimentation is determined through a group process in an open community meeting.
- Planning, based on information from various sources and on a clear objective for each experiment.
- Experimentation, when the CIAL implements the planned experiment.
- Evaluation, when the CIAL meets with the facilitator to evaluate collected data, draw conclusions and present results to the community.
- Analysis, in which the question, 'What have we learned?' is especially important, ensuring learning from both the process and the results.
- Feedback, for which regular, open meetings are held in the community where the CIAL presents results and expenditures. This ensures that their research products become public goods. The facilitator is responsible for feedback to the formal research system.
- Regional or national forums where CIALs can exchange information and results are held annually.

Reflections

Ecological interactions and local knowledge gaps are not always considered explicitly in the CIAL process, and no formal mechanism is incorporated to develop an understanding of the agroecological principles and interactions that may underlie the issues being researched. Such knowledge may be essential for the design of meaningful experiments. Methods for analysing the scale of action needed for successful intervention are not sufficiently developed. This can be limiting if the research issues are related to pests, diseases and many NRM situations. Relationships with the community depend strongly on the quality of social capital. In areas where trust is low and association along non-kinship lines is rare, forming and/or sustaining CIALs may be difficult. However, many advantages accrue. Farmers systematically evaluate technological options such that minimal risk is incurred. They learn basic research principles that provide for developing a common language and two-way communication with research and extension professionals. In some conflict-ridden communities CIALs have provided an entry point for building social capital. Although community-based, the radius of CIAL influence can be increased by forming networks of experimenting communities and creating second-order associations.

A recent impact assessment of the CIAL movement revealed that:

- Many CIALs provide an effective research service to their communities; they report their results publicly and cite widespread testing of these by local farmers.
- In some communities where farmers are applying CIAL recommendations, yields of staple crops have almost doubled. Food security can be greatly improved in CIAL communities. The poorest benefit the most from the increased availability of food during times of scarcity.
- More farmers in CIAL communities are experimenting with soil conservation practices and conduct experiments with a far greater diversity of varieties and crops than in communities without CIALs.
- Innovations identified by the CIALs reach local farmers more rapidly than others and also spread to other communities both with and without CIALs.
- CIALs report positive changes in the attitudes of the R&D professionals working with them.
- Women and marginalized social groups gain social status and respect in their communities as a result of belonging to a CIAL.

The effect of the CIALs on their communities and on formal research services transcends dollars and cents. Nevertheless, CIAT estimates the return on the investment made in developing and applying the CIAL approach at 78 per cent.

The CIAL movement is still young, and its future evolution is uncertain. Properly managed, CIALs can deliver substantial growth and equity benefits. However, their effects on the sustainability of production are less predictable. By allowing adaptive research to be devolved to the farming community, CIALs cut the costs of formal research while increasing its impact.

5

Development of Community Wastelands and Forests

Conservation of biodiverstity and protection of forests are the other important components of community forestry. However there has been a lukewarm response to these activities in the past, mainly because of lack of awareness, inadequate benefits and lack of trust between the local communities and government agencies who own these lands. Hence the new strategy should find a suitable solution for these problems.

Development of forest lands: Rehabilitation of denuded forest lands and protection of natural forests are essential to conserve biodiversity. The forests influence the climate and soil moisture conditions and improve agricultural production. This aspect needs wider publicity in rural areas and the local people should be motivated to take active part in Joint Forest Management. The Village Forest Committees willing to protect the natural forests, should be encouraged to share the benefits in the form of fodder, fuel, timber and NWFP. These committees can also take up reforestation of degraded forest lands, aiming to conserve biodiversity and generate income. While facilitating natural regeneration, the local communities can also introduce fuel, timber and NWFP species of their choice to enhance the productivity on denuded forest lands.

Development of community lands: Lack of people's initiative was a serious problem in developing community wastelands, mainly due to lack of direct benefits. Earlier efforts to establish woodlots could not generate substantial income. Establishment of NWFP species such as tamarind (*Tamarindus indica*), Indian gooseberry (*Emblica officinalis*), neem (*Azadirachta indica*), Pongamia (*Derris indica*), *Madhuca spp*. and a wide range of medicinal species can generate substantial income, regularly for over 50-100 years. With the world's herbal medicine business likely to cross US $ 3 trillion during the next decade, it will be an excellent opportunity for the Asian farmers to enhance their income through cultivation of these species on community lands as well as on private wastelands.

Development of Mined Areas: Rehabilitation of mined areas is another cause of serious concern. Although the industries are obliged to establish a green cover after mining the area, the outcome has been very poor. Such sites are major causes of soil erosion and flooding of rivers. Lack of commitment, inadequate expertise in selection of plant species and their management have been the main reasons for failure. These activities can be entrusted to voluntary organisations and local communities, with research support to raise healthy plantations. Preference should be given to NWFP and timber species which have long gestation period, sufficient to enrich the mined lands.

CO_2 Sinks: There are opportunities for environmental protection by establishing industrial greenbelts, recycling treated effluent and developing carbon dioxide sinks, particularly around urban areas. The polluting industries can be compelled to establish greenery through the involvement of voluntary organisations and local people.

Environmental Awareness and Eco-tourism: Lack of awareness among the common public is a cause for environmental pollution in the developing countries. Hence, continuous efforts are needed to motivate the common public to plant more trees and protect the environment. School children and literate urban population are fairly receptive to participate in environmental protection. This can be undertaken through school based plant nurseries, development of arboretum, botanical gardens and greenbelts around the schools and industries.

While developing industrial greenbelts and afforestation on river banks, special efforts can be made to promote eco-tourism. Creating waterbodies through plugging of gullies, developing, nature trails, shady grounds for resting, selecting suitable plant species to attract wildlife and providing adequate eco-tourism. Environmental exhibitions at these sites can help in motivating the public to take active part in community forestry and environmental protection.

ORGANISATIONAL DEVELOPMENT

Development of local organisations is the key to ensure people's participation and to sustain the benefits of investments and efforts made in community forestry.

TREE GROWERS ORGANISATIONS

The initial need is to provide leadership to organise local people in afforestation. This can be carried out through formation of local Selp-Help Groups, village level farmers' associations and forest groups. This activity should receive priority during the next decade.

Capacity Building

Training of local farmers, supply of critical inputs, credit and technical advisory services to improve the profitability and developing market networks are the critical inputs which have to be introduced in the field to support tree growers.

Networking of Tree Growers and User Groups

There is an urgent need to set up networks of tree growers, forestry users and concerned scientists in these fields. Such networks should identify the opportunities for tree growers. They can also promote processing and marketing of forestry products and protect the interests of tree growers.

Forestry Information Service

Setting up of information centres to identify the priority of the industries and farmers and provide necessary information to enhance the economic viability of community forestry would help to sustain the interest of tree growers. Forestry Extension Units should closely associate with Information Service Centres to identify appropriate technologies for transferring to the field.

Forestry Research

The networks should identify the problems affecting the production and marketing of the produce and encourage the Forestry Research Institutions to take up relevant scientific studies to solve the problems. The researchers should closely associate with the network and Information Centres to disseminate new technologies to solve field problems and enhance the production.

Forestry research should focus on the following topics:

- Conservation of biodiversity;
- Improving the productivity of wastelands, tree plantations and natural forests;
- Exploring the utility of multipurpose tree species and NWFP;
- Developing of alternatives for conserving wood;
- Selecting various organisational systems to manage community forestry.

The outcome of these research topics can further strengthen the community forestry programme in the country.

DEVELOPMENT OF FOREST PLANTATIONS

By the seventeenth century the decline in the area of native forest in European countries led to the planting of trees, largely to provide alternative sources of timber supply. Sometimes plantations were established for the provision of services or other products than timber. They might be planted as shelterbelts, for dune stabilisation, for amenity or for the supply of firewood. Plantations of native species were established at first in areas where forests occurred naturally, for example France, Germany, England and Scotland, but later also on previously unforested land. In France, for example, planting of

Pinus pinaster was started on the sand dunes of the Landes and with *Pinus sylvestris* and *Picea abies* on former agricultural land in the Vosges; in Germany likewise agricultural land was planted with Norway spruce *(Picea abies)* in Saxony. In tropical countries such as Myanmar (then Burma), teak *(Tectona grandis)* was planted as a native species in the *taungya* system.

Teak was one of the first exotic forest plantation species to be used, being planted in Sri Lanka and the island of Java (Indonesia) from early in the nineteenth century. Exotic species, such as Douglas fir *(Pseudotsuga menziesii),* Sitka spruce *(Picea sitchensis),* Japanese larch *(Larix kaempferi),* lodgepole pine *(Pinus contorta)* and poplar *(Populus deltoides),* were introduced to Europe during the nineteenth century, when they began to play an increasingly important role in forest plantation programmes. Other exotic species, particularly the eucalypts *(Eucalyptus* spp.) and wattle *(Acacia* spp.) from Australia, were introduced as exotics in tropical and subtropical countries from the middle of the nineteenth century, while *Pinus radiata* was introduced from California to New Zealand and other countries such as Chile from the early years of the twentieth century. Box 1.3 describes in detail the experience with forest plantations in Denmark.

The main reason for the establishment of plantations remained the decline in natural forest area and a scarcity of wood and, the conversion of woods and forests to agricultural and grasing use. In recent years increased areas of natural forest have been managed for nature conservation, recreation, wildlife parks, etc., and commercial wood production has been either reduced or eliminated. For example, over 25 000 ha of high-yielding hybrid poplar plantations have been established in the north-western USA between 1992 and 1997 in response to both increased demand for poplar wood for orientated-strand board and decreased supply from public forests. On the other hand, some new areas have become available in European countries as land is taken out of agriculture due to trade and market considerations.

OBJECTIVES

Forest plantations are tree crops that are in some, but not all, ways analogous to agricultural crops. They often have a simple structure, at least in youth, and are usually composed of one or a few species (but not varieties as in agriculture, except in a few cases such as the intensively bred poplars) chosen for their fast growth, yield of specified products and ease of management. When established for wood production they have a higher productivity of usable wood than natural forests, but due to the way they are managed they do not, indeed cannot, provide the full range of goods and services that natural, seminatural or even secondary forest can provide. Although the objective of many plantations is the production of industrial roundwood and/or fuelwood, many are established for environmental protection or other services (*e.g.* soil and water conservation, enhancement of agricultural production in agroforestry systems, carbon sequestration) and some also provide non-wood forest products, such as fodder, various foodstuffs, medicines, etc.

Area

The area of forest plantations throughout the world started to increase in the 1970s as many governments became concerned about wood supplies for industry, and in developing countries fuelwood supplies, and has continued to increase since. However, there are no reliable global figures for plantation areas because forests of native species in several developed countries in the temperate and boreal regions, especially in continental Europe, are frequently regenerated naturally and it is not possible to distinguish those areas where supplementary artificial planting has been done. Furthermore many countries consider their plantations as 'seminatural' forests over a certain age, because as stands mature the clear initial row layout of the trees is lost and other species naturally regenerate under the canopy. For example, Austria, Czech Republic and Finland, in responding to a questionnaire to collect data for the ECE/FAO Temperate and Boreal Forest Resource Assessment (TBFRA), stated that they had no plantations in their countries as defined by the TBFRA process.

In 1995, an approximate estimate of the area of plantations in developed countries was 60 million ha, comprising 13.7 million ha in North America, 22.2 million ha in the Commonwealth of Independent States, 12.1 million ha in Europe and 13.2 million ha in Oceania (Australia, New Zealand, Japan). The most significant areas of plantations were in the Russian Federation (17.3 million ha, 2.1 per cent of the country's total forest area), the USA (13.7 million ha, 6.3 per cent), Japan (10.7 million ha, 44.4 per cent), Ukraine (4.4 million ha, 46.8 per cent) (FAO/ECE 1996), Spain (1.9 million ha, 14.9 per cent), New Zealand (1.54 million ha, 19 per cent of the total forest area in 1996, 91 per cent of which was *Pinus radiata)* and Australia (1.04 million ha in 1994, 85 per cent of which was softwood species, mainly *P. radiata).*

The estimated 'net' plantation area of 55 million ha in developing countries in 1995 was about 2.8 per cent of the total area of forests in developing countries. In 1980 the net plantation area was assessed at about 40 million ha. It has thus increased by about 15 million ha in 15 years, but even this figure may be liable to error. Many developing countries, especially those with large forest plantation programmes, provided updated information on their present plantation plans to FAO in 1996 and 1997 from which the reported annual rate of new plantations of 3 million ha was derived. Note, however, that this is the reported or planned rate, which may not necessarily have been achieved. Most of the countries with large plantation estates indicated that they intended to double their plantation areas between 1995 and 2010.

It was estimated from reported figures that 57 per cent of the forest plantation area consisted of hardwood species and 63 per cent was established for industrial purposes. Nearly three-quarters of these plantations were in the Asia-Pacific region, where China (21 million ha) and India (20 million ha) dominate, while about 15 per cent were in Latin America and 10 per cent in Africa.

One of the trends in the tropics has been that the proportion of industrial plantations established in large blocks fell from 40 per cent of the total plantation area in 1980 to 35 per cent in 1990. The proportion of smaller plantations established through farm forestry or agroforestry programmes grew in importance during the period 1980-95, particularly in the Asia-Oceania region. Unfortunately, the figures on small-scale private or community-owned forest plantations are even less reliable than for large-scale plantations. Some of these farm forestry or agroforestry plantations supply industrial wood markets for pulpwood (*e.g.* Brazil and Thailand) or peeler logs (*e.g.* India). In many countries, particularly those with limited forest area, planted trees grown outside the formal forest area often provide the bulk of fuelwood, poles, construction wood, utility wood, as well as fodder and other non-wood forest products for household use. Of the area of hardwood plantations planted for industrial use, 30 per cent or nearly 10 million ha consists of eucalyptus, followed by acacias (3.9 million ha or about 12 per cent of the hardwood area) and teak (about 7 per cent). Short rotation plantations of hardwood species such as the eucalypts, acacias and *Gmelina arborea* have been grown for many years by the private sector, but the establishment of plantations of teak or other valuable hardwood species has been carried out only by government forest services because of their slow growth and hence delayed returns. However, the likelihood of reduced supplies of high-quality hardwood logs derived from natural forests, combined with increasing purchasing power and expected higher prices for logs, is leading to increasing interest in investment by the private sector in valuable hardwood species, especially teak, in a number of countries, for example India, Malaysia, Costa Rica and Ghana.

Of the softwoods grown for industrial purposes, fast-growing pines such as *Pinus radiata, P. patala and P. caribaea* constitute about 25 per cent of the area while other, often slow-growing pines (*e.g. P. kesiya, P. massoniana, P. merkusii, P. rox-burghii, P. halepensis, P. pinaster* and *P. wallichi-ana)* make up about 36 per cent.

Responsibility for the monitoring and regulation of plantation crops grown for food and certain other purposes has long been the job of the agricultural sector. In recent years, however, several of these crops (the main ones being rubber, coconut and oil palm) have been providing 'forest' products that have been used for wood and fibre. For example, rubber wood is now used for the manufacture of about 80 per cent of the furniture made in Malaysia, while coconut and oil palm trunks and the branches of rubber wood are used for various forms of reconstituted 'wood'. Rubber wood and coconut stems are derived from the conversion of old plantations formerly disposed of by burning, while oil palm fruit residues are used for medium-density fibre board. The development of these new markets has thus not only improved financial returns but also used the resources in an environmentally friendly manner.

The area of these species appears to have increased from the 14 million ha reported in 1990. The increase may be due to better coverage of the data, although it is known that oil palm areas are increasing rapidly, rubber tree areas are also increasing and coconut plantations are decreasing. Coconut plantations comprise the largest area (about 42 per cent of the total), rubber 36 per cent and oil palm 22 per cent. Most of the coconut plantations are in Indonesia (33 per cent of the area) and the Philippines (28 per cent); most of the rubber plantations are in Indonesia (34 per cent), Thailand (20 per cent) and Malaysia (18 per cent), while most of the oil palm plantations are in Malaysia (44 per cent) and India (29 per cent).

CONIFEROUS FORESTS AND TEMPERATE BROADLEAF

Of closed canopy forests, perhaps none has been as extensively utilised and altered by humans as the temperate forests. Significant portions of Europe and eastern Asia that once supported forests have long since been converted to pasture and agriculture, and little of the original vegetation remains. Similarly, much of the temperate forest in the USA has been cut, although a large percentage returned to forest after cutting or agricultural abandonment. None the less, forests still occupy moderate to significant portions of the temperate regions of North and South America, Europe, Asia and Australia, and are of significant economic and ecological importance.

Distribution and Extent

There are about 1.1 million km^2 of temperate broadleaf deciduous forests in the world, with perhaps another 3-8 million km^2 of temperate coniferous and broadleaf evergreen forests and woodlands. In North America, temperate broadleaf forests are confined primarily to the eastern USA and south-east Canada.

West of the central prairie region are the interior and mountainous coniferous forests, and beyond them the coastal temperate coniferous rain forests of the Pacific North-west. Latitudinally, they occur between the subtropical zone (about 28-30°N) and the boreal zone (46-47° N), although the temperate coniferous rain forests of the Pacific coast of Canada and Alaska grow a little north of 56° N. Smaller areas of temperate forest occur in southern Mexico and into Belize and Guatemala, principally at high elevations and along river courses at 0-2500 m.

The South American temperate forests grow primarily along the far south-western coast and the southernmost tip of the continent. The deciduous forests east of the Andes grow south of 45° S, while the coastal rain forests grow between 41 and 56° S. Temperate broadleaf evergreen forests grow principally in Chile, with a minor extension into Argentina. European temperate forests occur predominantly in western and central Europe, continuing in a narrow band through eastern Europe to the Ural Mountains in Russia. Temperate deciduous forests also occur in the Near East in a band around the southern Caspian Sea, broadening to the east into the Iranian highlands, Turkmenistan, Uzbekistan and Kazakhstan. Mountainous areas are forested primarily by conifers.

Asian temperate deciduous forests are found primarily in the central and south-eastern region, including mainland China, Taiwan, Korea and Japan. The region lies between about 26° and 50° N, with conifers increasing in importance with latitude and elevation. Temperate forests, primarily coniferous but often with abundant deciduous canopy species, also occur in central Asia, principally at higher elevations in the Himalayan region. In Australia, moist temperate deciduous forests are found predominantly in northern and western Tasmania, while the mainland is occupied principally by the evergreen *Eucalyptus* forests of the south-east and southwest. Much of New Zealand is temperate broadleaf evergreen forest.

Soils

Soils are naturally variable, although the dominant soil orders found under broadleaf temperate forests are Alfisols, Inceptisols and Ultisols, with substantial areas of Entisols in floodplain environments (Archibold 1995). Spodosols and Ultisols are most common under temperate coniferous forest, but ash-derived Andosols are common in volcanic regions. Alfisols, Inceptisols and some Ultisols are relatively fertile, with high base saturation. Soil pH typically ranges from slightly acidic to slightly basic, although under conifers and some hardwoods, or in low-lying areas with impeded drainage and decomposition, soil pH may be 4 or less.

Climate

As expected for a forest type distributed so widely, prevailing climatic conditions vary substantially. Generally, though, temperate environments are characterised by warm, relatively humid summers and cool-cold winters. Where precipitation is insufficient to support closed canopy forests, temperate forests variously grade into grassland (steppe or prairie), savannah or open woodlands. To the north, where both temperature and precipitation decline, they merge with the boreal forests, and to the south with (sub)tropical vegetation.

At lower elevations, mean temperature during the coldest month typically ranges between -5 and 10°C, with mean temperatures between 10 and 18°C. In mountainous regions, the temperatures can be significantly lower, approaching those of the more northerly boreal zone. Mean annual temperatures typically range between 8 and 13°C but can be lower or higher, depending on geographical location and local elevation.

Precipitation is year-round, although there are significant geographical differences in seasonal distribution. Winters are wetter than summers in much of the temperate forest region, although precipitation is more evenly distributed in portions of eastern North America, western Europe, far southern South America and New Zealand. In the temperate coniferous rain forests of the north-western USA and southeastern Canada, summers are dry and most precipitation falls during the late autumn and winter. Mean annual precipitation is less than 500 mm in some portions of South America and Russia. At the other extreme are the temperate coniferous rain forests of the north-west North American Pacific coast (cover photo), where annual precipitation approaches 2500 mm. More typical values for the temperate forests of eastern North America, western Europe and South-East Asia are 550-1300 mm.

Regional Formations

Although there are distinct regional compositional differences, there is remarkable taxonomic similarity among the Asian, European and North American formations. Numerous genera are common to all three, including *Acer, Betula, Fagus, Fraxinus, Populus, Quercus* and *Ulmus.* The flora of North America and eastern Asia are most similar to one another, and share several genera of forest trees not found in Europe, including *Catalpa, Diospyros, Liriodendron, Nyssa* and *Sassafras.* There are further such affinities in the herbaceous flora of the forest floor (Li 1972).

Although the modern floras evolved under similar climatic regimes, the taxonomic similarity is due primarily to their common geological past. Both the North American and Eurasian continents were connected until the mid-Tertiary, after the period (Cretaceous) during which most of the genera common to the temperate forests of the Northern Hemisphere differentiated. The South American and Australasian formations do not share a common geological past and differ tax-onomically from those of the Northern Hemisphere. However, they do share some families (*e.g.* Fagaceae). Of the three Northern Hemisphere formations, the Asian is the richest, followed by the North American and the European distantly third. The comparatively depauperate European temperate flora is believed to have resulted largely from the east-west orientation of the region's mountain ranges, which inhibited migration and reduced available refugia during past glaciations.

Asian

On the Asian mainland, temperate deciduous forests occupy a coastal band along the China Sea from the Korean Peninsula and into China in the vicinity of Tianjin. From here, deciduous forests broaden throughout the Chinese lowlands to the west (to the Mongshan Mountains) and south to the Yangtze River. Intermixed along the Chinese coast and in the interior highlands are mixed broadleaf-coniferous forests. Further to the south, evergreen broadleaf trees become more abundant. Coniferous forests become more prominent in the northern parts of both China and Korea. Japan's forests consist primarily of temperate broadleaf deciduous communities in the central-north region, with conifers more abundant at higher elevations and latitudes. To the south, between about 26° 30' and 35° 30'N, lie the temperate broadleaf evergreen forests. Ching (1991) has described the composition of the region's forest in some detail and I only briefly summarise them here.

China. In China's north-eastern temperate forest region (eastern Heilongjiang and Liaoning provinces), conifers occur at the highest elevations and the northern latitudes, lending the forests a (sub)boreal character. Below these forests (up to 1100m elevation) are the *Pinus koraiensis-mixed* hardwood forests, some of which are rather species-rich. Subtypes include the mixed hardwood-*Pinus koraiensis* forest, oak forests *(Quercus mongolica)* and *Betula* forests, the latter of which occur at higher latitudes and elevations.

China's northern temperate forest is roughly triangular in shape, extending from 32° 30' to 42° 30'N and from 103° 30' to 124° 10'E. The major types within the region include the northern oak forest and the southern oak forest. Common throughout the northern oak forest is *Quercus liaotungensis,* with increasing *Quercus mongolica* to the east.

The Liaotung hills and plains comprise the central portion of the northern oak forest area, with forest composition varying by landform and geography. In the mountainous regions of the Hebei and Laioning hills, broadleaf trees grow to elevations of about 1600 m, the forests being dominated by *Quercus* spp. along with *Betula* and *Populus.* The forest of the high loess plateaus of Shaanxi and Shanxi consist of *Pinus* and various hardwoods. The *Picea meyeri* and *Picea wilsonii* characteristic of the original primary forest have been widely replaced on the plateau by *Betula* in disturbed areas.

The southern oak forest of the northern region includes not only deciduous species but also broadleaf evergreens. It is distinguished from the northern oak forest by the scarcity of *Quercus liaotungensis* and the presence of the evergreen *Quercus variabilis.*

The vegetation of eastern Shandong bears close similarity to many of Japan's forests and is floristically diverse. Frost-tolerant *Tilia* spp. and *Corylus* spp. are found to the north, and *Machilus thunbergii* is present further south. The central and southern Shandong hills are forested with *Quercus*-dominated communities, as are the plains and hills of southern Shanxi and central Shaanxi. *Betula* and *Salix* become progressively more important with elevation, as do the conifers *Pinus armandii, Pinus tabulaeformis* and *Abies fargesii.*

The Yangtze River drainage in east central China is a transitional zone between the temperate forests to the north and the subtropical forests to the south. There is a high degree of endemism. The genus *Quercus* is well represented, and the tree flora is a mix of more northerly and southern species. Deciduous species and conifers are intermixed with the broadleaf evergreens, which include the genera *Actinodaphne, Castanopsis* and *Lindera.* Bamboo 'forests' grow here, including *Aruninaria* spp. and *Phyllostachus* spp. The upper river valley is particularly diverse, and includes 50 broadleaf and 12 coniferous genera.

Forest of *Abies, Picea* and *Tsuga* occupy the high elevations.

Korean Peninsula. There is disagreement over the classification and latitudinal zonation of Korea's forests but the following types are relatively distinctive: warm temperate broadleaf evergreen, southern deciduous, northern deciduous and cool temperate/boreal coniferous.

The warm temperate broadleaf evergreen forests occupy the southernmost coast of the Korean Peninsula, and include *Quercus* spp., *Castanopsis* spp., *Cinnamomum camphora* and *Machilus thunbergii.* Temperate deciduous forests lie north of approximately 35° N and account for about 85 per cent of Korea's forested land base. Species common to the southern deciduous forests are *Acer formo-sum, Carpinus laxiflora, Carpinus tschonoskii, Quercus mongolica* and *Quercus serrata.* The northern deciduous species include *Celtis koraiensis, Choseia bractesa, Hemiptera dividii, Larix olgensis, Magnolia parviflora* and *Populus maximowiczi.* Other species found variously in the temperate region include *Acer mono, Betula* spp., *Carpinus* spp., *Fraxinus* spp. and numerous *Quercus* spp. In the plateaus and mountains of northern Korea, coniferous species dominate the overstorey, including *Abies nephrolepis, Abies holophylla, Larix dahurica* and *Picea jezoensis. Pinus pumila,* a dwarf species, grows at the highest elevations (above 1200-1600 m). Broadleaf understorey species include *Acer* spp., *Betula* spp., *Tilia* spp. and *Ulmus* spp.

Japan. The forests of Japan include warm temperate broadleaf evergreen, temperate deciduous and cool/cold coniferous. The warm temperate broadleaf evergreen forests occur on the islands of Honshu, Shikoku and Kyushu, and extend from the lowlands to 600-850m. Species characteristic of the region include *Castanopsis cuspidata, Cycobalanus* spp., *Cinnamomum* spp., *Machilus* spp. and *Quercus* spp. In more mature forests, *Quercus gilva* grows on deeper soils, *Quercus slicina* on steep slopes and *Quercus glauca* over limestone. Following disturbance, *Quercus acutissima, Quercus serrata* and *Quercus variabilis* establish in mid-succession. The conifers *Abies firma, Podocarpus macrophyllus, Podocarpus nagi, Torreya nucifera, Pinus thunbergii* and *Tsuga sieboldii* variously occur, depending on latitude and proximity to the coast.

Temperate deciduous forests on the four main Japanese islands (Hokkaido, Honshu, Shikoku and Kyushu) are diverse in species and habitat, too much so to adequately describe here. In general, though, *Fagus crenata* dominates many forests. Its associates vary, depending on topography, disturbance history, soils, etc. Sites facing the Sea of Japan are commonly cold, snowy and windy, and species like *Acer tschonoskii, Acer matsumarae, Alnus maximowiczii, Quercus mongolica* and *Sorbus commixta* are common.

Various species of *Alnus, Populus* and *Salix* grow in floodplain environments, and the conifers *Abies sachalinensis, Abies firma, Abies mariesii, Picea jezoensis, Pinus thunbergii* and *Tsuga sieboldii* variously grow with species of *Betula, Carpinus, Fraxinus, Magnolia, Quercus, Tilia* and *Ulmus* in montane regions. *Quercus cripsula* is also a common dominant, particularly on Hokkaido, and grows well on somewhat drier sites than *Fagus crenata.* Its associates include *Abies sachalinensis, Acer mono, Betula ermanii, Kalopanax pictus, Ostrya japonica, Picea jezoensis, Tilia japonica* and *Ulmus davidiana* var. *japonica. Pinus densiflora* sometimes forms pure stands, with a broadleaf understorey. At higher latitudes and altitudes, the cool/cold coniferous forests are dominated by *Picea* spp. and *Abies* spp. Broadleaf associates include *Betula* spp., *Fraxinus mandshurica, Kalopanax septemlobus* and *Fagus crenata.*

Australasian

Australia and Tasmania. The majority of Australia's forests are considered (sub)tropical, some of which (the south-western and eastern *Eucalyptus* forests) have been described above. Webb and Tracey (1994) consider Australian forests south of about 38° S to be temperate, and I adopt that latitudinal boundary here. However, others consider the temperate forest zone to extend as far north as about 25° S, and in reality the transition from (sub)tropical to temperate is characteristically indistinct.

Regardless of the precise line of demarcation, the temperate forests of the Australian mainland are dominated primarily by *Eucalyptus,* and account for approximately 29 million ha of a total 42 million ha of closed forest, both temperate and tropical. Groves (1994) has described the dominant vegetation in some wet sclerophyll Australian temperate *Eucalyptus* forests distributed among four main geographical regions. Similar tabulations, including both wet and dry sclerophyllous *Eucalyptus* forests divided into major alliances, have been made by Ovington and Pryor (1983). Despite the diversity of *Eucalyptus,* temperate forests are often dominated by only a few species, for example *E. obliqua-E. viminalis* forests in parts of Victoria and *E. marginata-E. calophylla* forests in Western Australia.

Temperate rain forest occupies small areas of south-eastern Australia, principally in the territories of Victoria, the north-eastern corner of New South Wales and far south-eastern Queensland. The canopies of Victorian forests are dominated by *Nothofagus cunninghamii,* while *Athersperma moschatum, Acacia melanoxylon, Eucriphia lucida* and *Phyllocladus asplenfolius* are occasional or common associates. *Nothofagus cunninghamii* does not grow in easternmost Victoria, and *Athersperma moschatum, Eleaocarpus holopetalus* and *Telopea oreades* are most common. Eucalypts dominate the drier uplands in the region. In the forests of New South Wales, *Eucryphia moorei, Nothofagus moorei* and *Doryphora sassafras* are common, with the former two species occurring at somewhat higher elevations.

Western Tasmanian temperate rain forests are dominated by *Nothofagus cunninghamii,* which may contain any of about 70 associates, only a few of which typically occur together in a stand. *Athrotaxis* spp. (especially *A. cupressoides)* occur at altitudes greater than 900 m. In drier environments, like eastern Tasmania, *Eucalyptus* spp. are dominant; common ones include *E. delegatensis, E. dalrym-pleana, E. gunnii, E. archeri, E. coccifera, E. urnig-era* and *E. pauciflora.*

New Zealand. The climate of New Zealand ranges from subtropical to cool/ cold temperate, but here I consider it mainly temperate. Generally, the native taxa are similar to those of the South American and Tasmanian forests. Most are evergreen in habit, including both conifers and broadleaf species. There are two broad categories of New Zealand forest: conifer-broadleaf, and beech.

The conifer-broadleaf forests are predominantly lowland and are dominated by podocarps, including the genera *Dacrycarpus, Dacrydium, Phyllocladus* and, of course, *Podocarpus.* Also present are two species of *Libocedrus* and the kauri *(Agathis australis).* The warmer inland forests support primarily *Podocarpus spicatus* and *Podocarpus totara* as canopy dominants, along with a more minor component of *Podocarpus dacrydiodes* and *Dacridium cupressinum.* Hardwood associates of lower stature include the genera *Beilschmiedia, Knightia, Laurelia, Litsea* and *Nestegis.* Forests range from relatively open to closed canopy, depending on species and habitat. For example, *Dacridium cupressinum, Podocarpus hallii* and *Podocarpus totara* are most common on drier sites and may form near-monospecific stands, while *Podocarpus dacrydiodes* often dominates on wetter, swampy sites. In the coastal hills and plains, *Dacridium cupressinum* is commonly dominant, although *Podocarpus hallii* and *Podocarpus totara* are likely to occupy more xeric sites.

Genera common and often dominant in coastal lowland forests include *Elaeocarpus, Met-rosideros* and *Weinmannia*. The conifer-broadleaf shrub community in these forests is often well developed, with abundant species. The overstorey tree *Weinmannia racemosa* is abundant and widely distributed on the wetter, western side of both islands. Another important tree species is the giant kauri *(Agathis australis),* much reduced in extent and growing primarily north of 38° S, particularly on ridges. It grows with podocarps and various hardwoods, and occurs both singly and in groups or thickets on poorer soils.

The beech, or *Nothofagus,* forests occupy cold, wet, montane environments and are characteristically lower in stature and less species-rich than the conifer-broadleaf forests. Four species of *Nothofagus* grow in New Zealand: *N. solandri, N. menziesii, N. fusca* and *N. trun-cata. Nothfagus solandri* is the most widespread of the four species and there are two recognised varieties: var. *solandri* and var. *cliffortoides.*

The former occupies warm, dry lowland environments primarily from East Cape on North Island to the centre of South Island, while var. *cliffortoides* grows at higher altitudes or on poor soils under high rainfall to the south. *Nothfagus solandri* forms nearly pure stands, with a sparse under-storey composed of shrubs (especially *Coprosma* spp.), lichens and mosses, although both varieties co-occur where their geographical ranges overlap. *Nothofagus fusca* is distributed from East Cape to the southern end of South Island, and grows best on deeper, well-drained soils. *Nothofagus menziesii* has similar site requirements and dominates the subalpine zone of western South Island and the eastern mountain ranges of North Island. *Nothofagus truncata* grows on drier, less fertile sites, primarily north of 42° S.

Near Eastern

The Near Eastern temperate forest lies between 35 and 45° N, from about 27° E in eastern Turkey to about 58° E in eastern Iran. The forests of this region are a mixture of both European and Colchian species, with elements of (sub)Mediterranean and steppe flora.

In western Turkey, oak forests are common, composed of various species of *Quercus* and other hardwoods. *Fagus orientalis* grows at higher elevations, though not in abundance. In the western portion of the Anatolian plateau in Turkey, *Quercus*-dominated temperate broadleaf forests grow between 1000 and 2000 m and contain a number of other genera as well. Extending into Iran further to the east are other forests dominated by *Quercus.* Various species of *Acer, Pyrus* and *Prunus* also occur, as well as a few species, like *Fraxinus excelsior,* more commonly found in the Middle European and Submediterranean regions. On alluvial soils at lower elevations (<400m) grow more diverse forests, which in addition to *Quercus* spp. include *Alnus barbata* and *Carpinus* spp. Moist forests between the Caspian Sea and Black Sea contain a number of species, including many missing from the European forests to the west. Further east, the species and assemblages change but a few genera predominate, with a number of low-stature woody species occupying the understorey (*e.g. Buxus, Crataegus, Corylus, Ilex, Ligustrum, Rhododendron* and *Sambucus).*

At high elevations (600-1100 m) are the beech *(Fagus orientalis)* forests. Near pure *Fagus orientalis* forests grow between the Sakarya River in western Turkey and the Turkey-Georgia border. Between 600 and 1300 m on calcareous northern slopes on the mountains of the southern Crimean Peninsula, *Fagus orientalis* grows alone or with *Carpinus betulus, Fraxinus excelsior, Tilia cordata* and *Ulmus glabra.* On southern slopes, it is found with *Acer hycranum, Pinus nigra, Carpinus orientalis* and *Quercus petraea.* In the Caucasus Mountains, it grows between 1000 and 1500m with *Acer* spp., *Fraxinus excelsior, Tilia platyphyllos* and *Ulmus glabra.* At higher elevations, *Fagus orientalis* grows with coniferous species like *Abies born-mülleriana, Abies equitrojani, Abies nordmanni-ana, Picea orientalis* and *Pinus sylvestris,* along with some of the broadleaf species found at the lower elevations.

North and Central American

Temperate forests occupy much of North America, from north of the border between the USA and Canada south into Mexico and some Central American countries at high altitudes. The conifer-dominated forests of western North America are separated from the more deciduous forests of the east by vast areas of former prairie now dominated by agriculture where rainfall or irrigation permit.

USA and Canada. The eastern North American temperate forest is primarily deciduous, although the forests of the south-eastern USA contain large amounts of conifers, both natural and planted, as do the states bordering the Great Lakes. The forests of the region have been extensively described, especially by Braun (1950) and more recently Barnes (1991). I follow Barnes's (1991) classification scheme, which is derived from Braun's (1950) and Küchler's (1964).

The South-eastern Evergreen region covers the coastal plain of the south-eastern USA, stretching roughly from Virginia along the Atlantic coast, continuing south across Florida and the Gulf of Mexico to Texas. *Pinus palustris-Aristida stricta* savannahs formerly dominated much of the uplands but have been largely converted to other uses or cover types, and both *Pinus elliottii* and *Pinus taeda* are presently the most abundant conifers in the region. *Magnolia grandiflora* and *Fagus grandifolia* are believed to be climax dominants on mesic sites, although a number of other genera are abundantly represented.

Inland of the South-eastern Evergreen region is the Oak-Hickory-Pine region, extending from Pennsylvania to Alabama and bordered by the coastal plain towards the Atlantic Ocean and the Appalachian mountains to the west. As its name implies, the principal species include those in the genera *Quercus* and *Carya,* often in mixture with species common to adjacent regions, including *Pinus.*

The Appalachian Oak region includes the Appalachian mountains, from Georgia north and eastward into eastern New York and Massachusetts, and is the most floristically diverse in North America. Oaks are abundant at all elevations and aspects. At higher elevations (> 1100 m), species are more characteristic of the beech-sugar maple and sugar maple-basswood forests to the north. At the highest elevations are boreal forests composed of *Picea rubens* and the endemic *Abies fraserii* in the south, and *Picea rubens* and *Abies balsamea* to east to south-west from western Pennsylvania to northern Alabama, and coincides physiographi-cally with much of the unglaciated Appalachian plateaus. The vegetation comprises principally species from the regions which it borders. The Western Mixed Mesophytic region lies immediately to its west, covering central and western Tennessee and Kentucky, the southern boundary of the Wisconsinan Glaciation in southern Indiana and Ohio, and northern Mississippi and Alabama. Forests dominated by *Quercus* spp. cover most of the region, with a large component of *Carya* and *Pinus.*

The westernmost region is the Oak-Hickory, bordered to the west formerly by prairies and presently by agricultural lands. It extends from Manitoba, Canada to southern Texas. South of the glaciated portion lie the Ouachita Mountains and Ozark Plateau. *Quercus alba* is the dominant species on dry-mesic uplands, with various species of *Carya, Quercus* and *Pinus* characteristic of drier sites. In the glaciated region to the north, *Quercus* spp. are common on drier uplands, while *Fagus grandifolia* and *Acer sac-charum* grow on more mesic sites. *Quercus macrocarpa* formerly formed savannahs interspersed with prairie, most of which has been lost to agriculture and urbanisation.

The Sugar Maple-Beech region occurs on glaciated till in the southern half of Michigan's lower peninsula, Indiana and Ohio; *Acer saccha-rum* and *Fagus grandifolia* are the dominant species in mature forests, although a number of other species are common. The Sugar Maple-Basswood region to the north of the Sugar Maple-Beech region is situated on glaciated terrain and, as the name implies, *Acer saccharum* and *Tilia americana* are dominant. Other associates are those characteristic of the Sugar Maple-Beech region.

The Northern Hardwood-Conifer region extends across the northern USA and southern Canada from Manitoba to Maine and Nova Scotia. The terrain is glaciated and supports a wide array of forest types. Species common to the Sugar Maple-Beech region predominate on glacial till, although the region extends beyond the northern range of *Carya, Liriodendron, Nyssa* and other genera common to the more southern forest regions. *Quercus ellipsoidalis* grows on dry glacial outwash plains, but fire-prone *Pinus banksiana* is the most dominant species on these soils. Much of the presettlement forest was dominated by *Pinus strobus* but virtually all of the original forest was cut and converted naturally to hardwoods, including *Betula papyrifera, Populus tremuloides, Populus grandidentata* and *Quercus* spp. Coniferous swamps on calcareous soils often consist predominantly of *Thuja occidentalis* with lesser amounts of other hardwoods.

The temperate forests of western North America differ from those in the east by a preponderance of conifers rather than hardwoods. Furthermore, much of forested western North America is mountainous and the vegetation is often distributed in rather discrete elevational bands, as opposed to the more subtle changes with landscape position associated with the eastern deciduous forests.

The Rocky Mountains extend from north-central Canada, through the USA and into Mexico, and form the eastern border of North America's temperate coniferous forests. At the lowest elevations (1500-1600m), *Pinus edulis* and *Pinus monophylla* grow with various species of *Juniperus* and *Quercus.* At higher elevations are *Pinus ponderosa* forests, both closed canopy and open park-like stands, above which lie forests dominated by *Pseudotsuga menziesii.* In the southern Rockies, the associates of *Pseudotsuga menziesii* are few, but in the northern Rockies it is variously found with *Abies grandis, Larix occidentalis, Pinus contorta* and *Populus tremuloides.* With increasing elevation in the northern Rockies, *Pseudotsuga menziesii* occurs with *Thuja plicata* and *Tsuga heterophylla.* These two species often form nearly pure stands. *Populus tremuloides* grows at similar elevations, colonising after disturbances, principally fire. Just below the treeline are found *Picea engelmannii* and *Abies lasiocarpa.* In the far northern Canadian Rockies, the principal sub-alpine species are *Abies lasiocarpa, Picea glauca* and *Pinus contorta.*

In the Sierra Nevada Mountains of California, *Pinus sabiana* is a dominant conifer in the western foothills, growing with *Quercus* spp. On western slopes, *Pinus lambertiana* grows with *Pinus ponderosa* at higher elevations (750-1800 m), together with *Abies concolor* and *Calo-cedrus decurrens.*

Pinus jeffreyi occupies lower eastern slopes and higher west-facing elevations. Further south, *Pseudotsuga macrocarpa* and *Pinus coulteri* grow in the elevations between desert scrub and the mixed conifer forests described above. *Sequoiadendron giganteum,* the world's largest trees, are distributed between 1350 and 2250 m on the western slopes of the Sierra Nevada. *Abies magnifica* and *Pinus contorta* grow between roughly 1800 and 2400 m, in regions of extremely high snowfall (4-20 m snow-pack). The subalpine forests (>2400m) consist of *Pinus albicaulis, Tsuga mertensiana, Pinus balfouriana, Pinus flexilis* and the long-lived (>4000 years) *Pinus aristata.*

The coastal mountain ranges of the northwestern USA and south-west Canada differ from the interior mountains principally by the abundant precipitation they receive (upwards of 2000 mm annually) and the exceptionally large size that many of the species attain. Along the Pacific coast from northern California into southeastern Alaska and the Kodiak Islands, *Picea sitchensis* forests are found in narrow bands (generally only a few kilometres wide) below 150 to 600 m elevation. Its associates include *Pseudotsuga menziesii, Thuja plicata, Tsuga heterophylla* and *Alnus rubra.* In southern Oregon and northern California, *Sequoia sempervirens* forests grow at similar elevations, extending as far as 15-20 km inland, and growing on slopes and sometimes at higher elevations.

Above the *Picea sitchensis* zone is the *Tsuga heterophylla* zone (550-1200m), above which *Abies amabilis* dominates the elevational bands ranging between 600 and 1500 m. The highest elevation forests are the subalpine *Tsuga mertensiana* forests that grow as high as 2000 m in southern Oregon. Principal associates include *Abies amabilis, Abies lasio-carpa* and *Abies procera. Tsuga mertensiana* is replaced by *Abies lasiocarpa* in the interior portions of the Cascades and on the drier eastern slopes. The most extensive forests of the eastern slopes of the Cascades are those dominated by *Abies grandis,* which grow at elevations of 1000-1500m in the north and 1000-2000m in southern Oregon.

Central America and Mexico. Geographically speaking, the forests of the region are primarily (sub)tropical in nature, although there is a large degree of taxonomic similarity between the vegetation of the USA and Mexico and Central America. Some of the similarity is due to the continuity of species' range distributions, but it is also believed that the region provided a refuge during cool periods of the Oligocene, after which geological uplift and further climatic changes isolated the temperate forests to disjunct islands among otherwise (sub)tropical forests from Mexico southward into Nicaragua.

Among deciduous species, *Carpinus carolini-ana, Liquidambar styracifl.ua, Nyssa sylvatica, Ostrya virginiana* and *Prunus serotina* grow in both the USA and Mexico. Species with a high degree of similarity include *Fagus mexicana* and *Fagus grandifolia, Carya ovata* var. *mexicana* and *Carya ovata,* and *Acer sutchii* and *Acer saccha-rum.* Several genera are well represented in both regions, including *Alnus, Carya, Fraxinus, Juglans, Magnolia, Platanus, Populus* and *Salix.* The genus *Quercus* is especially abundant, both deciduous and evergreen.

In montane deciduous forests, *Juglans olancha, Liquidambar styraciflua* and *Ostrya virginiana* are dominant, growing at 1200-1800m in the Sierra de Chucharas in the Mexican state of Tamaulipas. Forests dominated by *Quercus* spp. and *Pinus* spp. often occupy the highest elevations (> 1800 m), with (sub)tropical species prominent in the lower elevations. Riverine forests may be composed primarily of deciduous genera like *Carpinus, Fraxinus, Platanus, Populus, Salix* and *Ulmus,* but uplands typically contain representatives from more (sub)tropical families like Clethraceae, Fabaceae, Sabiaceae, Lauraceae, Staphylaceae, Cunoniaceae and Rutaceae.

Mexico and Central America also contain a rich coniferous flora, especially of pines. There are at least 47 *Pinus* species, which is about half the world total. Many species are five-needled pines, and they occupy niches from coastal to high montane environments. Some, *e.g. Pinus caribaea* and *P. patula,* are widely planted in the tropics.

European

Temperate forests once covered much of Europe and the British Isles, but most of the original forest has long since been converted to agriculture. Moreover, many introduced species have been extensively planted, for example plantations of the western North American conifers *Picea sitchensis* and *Pseudotsuga menziesii* in the British Isles. None the less, temperate forests remain an important resource and there are numerous recognised community types. Jahn (1991) has extensively described the European deciduous forests, and I make considerable use of his excellent work in my more abbreviated summary here. Two regions of temperate forest can be recognised: the Middle European, which occupies a band between about 43° and 55-60°N, and the Submediterranean, which generally follows the southern border of the Middle European south to a latitude of about 40° N.

Middle European. In the Middle European region are four somewhat distinctive provinces: the Atlantic, Subatlantic, Central European and Sarmatic. Throughout the region are genera common to most of the temperate forest of the Northern Hemisphere, including *Acer, Betula, Carpinus, Fraxinus, Pinus, Populus, Prunus, Quercus, Sorbus, Tilia* and *Ulmus*. The genus *Fagus* is represented principally by *Fagus sylvatica,* which is distributed throughout most of the Middle European region and is, alongside *Quercus,* the most conspicuous and perhaps most ecologically important tree in Europe.

The Atlantic province includes the British Isles and the coastal regions of western Europe, and a number of characteristic native species are, or were, widely distributed. In the British Isles, *Quercus* spp. still dominate many sites, although *Acer* spp., *Carpinus betulus, Fraxinus excelsior, Tilia* spp., and other species are abundant. *Fraxinus excelsior* is often the dominant species on high pH soils, with *Quercus robur* and *Quercus petraea* on sites of lower pH and fertility. *Fagus sylvatica* and *Fagus sylvatica-Quercus petraea* woodlands occur in south-eastern England. *Fagus sylvatica* is also common in north-western France, portions of Belgium and throughout the Middle European and portions of the Submediterranean regions. In the Atlantic province, *Fagus sylvatica* is typically absent only on sites that are wet, steep, calcareous or very sandy. The Subatlantic province is orientated along a north-east to south-west axis, covering an east-west region between about 8 and 15°W in southern Fennoscandinavia and about 3° W and 3°E in southern France and northern Spain. *Quercus robur* is common in the north and co-occurs with a number of associates depending on soil conditions. *Fagus sylvatica* is prominent on upland sites in the north and in the hills and mountains to the south, and forms a complex regional mosaic of communities, determined by a combination of climatic, soil and physiographic factors. In the southern portion of the Subatlantic province, *Quercus* spp. become increasingly prominent, particularly at lower elevations. In the eastern Pyrenees, *Quercus pubescens* is found at the lowest elevations, *Fagus sylvatica* and *Abies alba* at higher elevations, with *Abies alba* increasing in importance with altitude.

The Central European province covers the former East Germany, western Poland and western Czech Republic and is characterised by a more continental climate than the Atlantic or Subatlantic provinces. *Fagus sylvatica* reaches its easternmost extent here, and occurs extensively north of the Baltic Sea. It has many associates, depending on elevation, and often has a more species-rich ground layer than comparable forests in the Subatlantic province. *Carpinus betulus-Quercus petraea* communities formerly occupied rich, mesic plains and uplands, but most have been converted to agriculture. *Betula-Quercus* forests grow on nutrient-poor, low pH soils, with *Betula pubescens* and *Quercus robur* common on wet sites and *Betula pendula* and *Quercus petraea* on drier sites. *Pinus sylvestris* and *Picea abies* are locally intermixed in the north and east, with *Larix decidua* common to the hills of southern Poland.

The Sarmatic province extends in a band eastward to nearly 60° E, its north-south boundaries tapering from about 50-60° N in the west to about 53-54° N at its eastern edge. To the south lies steppe and to the north boreal forests. Generally, the Sarmatic province is characterised by a progressive west-to-east loss of the deciduous species characteristic of the three provinces to the west and an increase in the importance of conifers. For example, there is a tendency for nutrient-poor sites to shift from domination by *Betula* and *Quercus robur* in the Subatlantic province to domination by *Pinus* in the Sarmatic. Similarly, on more fertile sites, *Fagus sylvatica* is no longer present, and has been replaced by *Tilia, Quercus robur, Carpinus* and *Pinus. Pinus sylvestris* is distributed throughout the Sarmatic province and is an associate or dominant on all but the most fertile, moist soils.

On sandy soils it forms nearly pure stands, without deciduous associates. *Picea abies* is also common, occurring with *Pinus sylvestris,* and increasing in importance with soil fertility and moisture. *Abies alba* occurs in the eastern part of the province, with *Abies sibirica* in the far western portion. *Populus tremula* and *Quercus robur* are present along the southern boundary, although the latter is uncommon in the north. In contrast to the northern portions of the Sarmatic province, the south is largely deciduous. *Quercus robur* and *Tilia cordata* are dominant, but *Carpinus betulus* is an important (though often low stature) associate.

Intermixed or adjacent to these four provinces of the Middle European region are the forests of the Alps and the Carpathian Mountains. The Western Alps (orientated north-south and running from the Mediterranean to Grenoble, France) are occupied by *Quercus pubescens* forests in the west-facing foothills, with *Fagus sylvatica, Quercus petraea* and *Pinus sylvestris* occurring at higher elevations. High-elevation interior forests contain a mix of deciduous and coniferous species. East-facing forests are predominantly deciduous hardwoods, with a large *Quercus* component.

In the Middle Alps, the north-facing foothills are forested with *Quercus petraea* and *Carpinus betulus,* which are replaced by *Fagus sylvatica* and *Acer pseudoplatanus* at higher elevations, *Abies alba* and *Picea abies* in the interior,

Fagus sylvatica on upper south-facing slopes and *Cas-tanea sativa* and *Quercus petraea* on the south-facing foothills. The Eastern Alps (Verona, Italy to the Dinara mountains of the former Yugoslavia) are occupied by a variety of hardwoods, with an increasing abundance of conifers at higher elevations. Similar species occur on the south-facing slopes, with the addition of *Fraxinus ornus, Ostrya carpinifolia* and *Quercus* spp. in the low foothills and foothill-plains transition.

The Carpathian Mountains extend in an arc from Slovakia east through south-west Ukraine and into central Romania. The forests of the foothills and submontane zone (up to about 600 m) of the western Carpathians are predominantly mixed hardwood forests, with a minor component of *Fagus sylvatica.* On the eastern side, forest and steppe alternate at the lowest elevations, and are progressively replaced by mixed hardwoods, and then coniferous forests, as elevation increases.

Submediterranean. The Submediterranean region is characterised by a preponderance of deciduous oaks *(Quercus* spp.) and represents the southern distribution of many of the species common to the mixed forests of the Middle European region to the north. To the south, where summer droughts are more extended, grow the evergreen Mediterranean forests. The vegetation is quite variable and complex throughout and is sensitive to climatic and site factors in this region, which borders temperate and Mediterranean climatic zones. Deciduous species of the Submediterranean region include *Carpinus orien-talis, Fagus sylvatica, Fagus moesica, Fagus orientalis, Fraxinus ornus, Ostrya carpinifolia, Quercus cerris, Quercus faginea, Quercus frainetto, Quercus pubescens* and *Quercus pyren-aica.* Various species of *Acer* and *Sorbus* are also present. *Fagus* spp. are characteristic of montane regions and co-occur with *Abies* spp. and *Juni-perus* spp.. *Pinus* spp. are present, with *Pinus nigra* growing with *Quercus cerris, Quercus faginea, Quercus pubescens* and *Quercus pyrenaica* in submontane environments, *Pinus sylvestris* common in montane zones, and *Pinus peuce, Pinus heldreichii* and *Pinus mugo* growing at high elevations.

South American

The South American temperate forests are the smallest in extent, confined primarily to Chile within and between the coastal and Andean mountain ranges, with a small extension into Argentina. Both broadleaf evergreen and deciduous species are present, often in mixture, although evergreen species are dominant. *Nothofagus* spp. are probably the most abundant trees; a total of 10 species grow in the region, of which seven are deciduous *(N. obliqua, N. alpina, N. pumilio, N. antarctica, N. glauca, N. leoni* and *N. alessandri)* and three evergreen *(N. betu-loides, N. dombeyi* and*N. nitida).* The only other deciduous species present is *Acacia cavens,* which grows in a more Mediterranean climate in the northern part of the temperate region, between about 31 and 38° S.

Nothofagus glauca and *N. obliqua* grow in the wetter climate to the south of the *Acacia* forests in central Chile, the former on eastern and northern slopes and the latter on lowland sites. *Nothofagus alessandri* mixes with *N. glauca* on moister sites of the coastal mountains. *Nothofagus obliqua* dominates much of the lowlands and valleys south of 36-38° S, depending on proximity from the ocean. It is often found with *N. alpina* and *N. dombeyi,* the latter of which often forms pure stands in cool wet gorges.

The conifer *Libocedrus chilensis* is often present on dry slopes opposite *N. obliqua,* and *Laurelia sempervirens* and *Persea lingue* are common, though lower-stature, associates of *N. obliqua.* At higher elevations in the Andes, *N. dombeyi* replaces *N. obliqua. Nothofagus alpina* forests occur in narrow elevational bands in the northern Chilean Andes and the species is widespread in mixed forests of the coastal range, often found with *N. dombeyi* and the associates of *N. obliqua. Auracaria araucana* occurs over small areas of the coastal range and is more widespread in the Andes (37-41° S) above the *N. alpina* forests (600-900m). *Nothofagus pumilo* occupies cold snowy sites, at elevations of 1500-1800 m at about 37° S down to sea level at Cape Horn. *Nothofagus antarctica* is a high-elevation krummholz-forming species that also colonises primary successional habitats (*e.g.* volcanic deposits) and other harsh or exposed sites. *Nothofagus nitida, Podocarpus nubigenus* and *Weinmannia trichosperma* occupy humid wetter sites between 600 and 900 m elevation south of about 40° S. *Fitzroya cupres-soides* is a large-sized tree occurring on wetter sites between about 40° and 43° 30'S in both the Andean and coastal ranges, and at elevations from about 700m to the timberline.

SUSTAINABLE LAND MANAGEMENT

In a wider context the term sustainable land management is used in soil and environmental protection, in preservation of ecosystem services and mineral extraction. Sustainable is also used in property and estate management as well as regional planning. Strictly speaking (in a wider /opposite context) the term "sustainable land management" is applied in forestry (this is where it comes from originally), agriculture, land surveying and in combination with land use management.

INCONSISTENT APPLICATION

In the international context the term sustainable is often used in the context of development services. In the course of their work organisations such as the United Nations (incl. sub-organisations and programmes *e.g.* UNEP / UNDP, FAO), World Bank, European Commission and development organisations make use of the term. In the centre of attention is the sustainable use of resources and issues around soil protection.

The World Bank defines Sustainable Land Management as a process in a charged environment between environmental protection and the guarantee claim of ecosystem services on the one hand. On the other hand it is about productivity of agriculture and forestry with respect to demographic growth and increasing pressure in land use.

"SLM is defined as a knowledge-based procedure that helps integrate land, water, biodiversity, and environmental management (including input and output externalities) to meet rising food and fiber demands while sustaining ecosystem services and livelihoods. SLM is necessary to meet the requirements of a growing population. Improper land management can lead to land degradation and a significant reduction in the productive and service (biodiversity niches, hydrology, carbon sequestration) functions of watersheds and landscapes."

The United Nations Economic Commission for Europe (UNECE) applies the term in a much wider context. Besides agriculture and forestry they include the mineral extraction sector, property and estate management.

"Land management is the process by which the resources of land are put to good effect. It covers all activities concerned with the management of land as a resource both from an environmental and from an economic perspective. It can include farming, mineral extraction, property and estate management, and the physical planning of towns and the countryside."

In the course of national politics and programmes only few European states use the terminology "sustainable land management". Here Australia and New Zealand are to be mentioned as both countries have agreed on sustainable land management with respect to climate change as part of their government programmes.

In the European context the definition of the European Network for Land Use Management for Sustainable European Cities (LUMASEC) may be used as a reference. It emphasizes the inter- and transdisciplinary cooperation on sustainable land management:

"As management is the human activity meaning the action of people working together in the aim to accomplish desired goals, land use management is a process of managing use and development of land, in which spatial, sector-oriented and temporary aspects of urban policy are coordinated. Resources of land are used for different purposes, which may produce conflicts and competitions, and land use management has to see those purposes in an integrated way. Therefore, land management covers the debate about norms and visions driving the policy-making, sector-based planning both in the strategic and more operative time spans, spatial integration of sectoral issues, decision-making, budgeting, implementation of plans and decisions and the monitoring of results and evaluation of impacts."

Research on Sustainable Land Management

Since 2010 various regional projects from science and praxis work on several topics regarding the sustainable use of land and its management. The research projects are embedded in the international research programme "sustainable land management" which is funded by the German Federal Ministry of Education and Research (BMBF).

As an example, a research on land management is carried out by a research team gathering researchers from Zambia, Lancaster and Birmingham universities. The different researchers are experts in biology, social behavior, land management,...The aim of the research is to help people in Zambia to have a good use of the Jatropha curcas tree using seeds and its efficiency to be a repellent for animals. It is a sort of sustainable land management. The project is called BKS: Bridging Knowledge system for pro poor management ecosystem services.

Land is a very valuable but limited resource, which in the future needs to be used in an intelligent way. Thus on the one hand the aim of the programme is to understand the interaction between land management, climate change and ecosystem services. On the other hand innovative system solutions will be developed to encounter the current challenges of global change and the accompanying land use conflicts.

COMMUNITY PROTESTS AGAINST MINING COMPANIES

Civil society and local communities in Asia have been actively protesting mining activities that cause water pollution and illnesses. The American-owned Newmont mining company attracted national attention in July 2004, when the people of Buyat Bay, off Sulawesi Island in Indonesia, demanded compensation for the disposal of tons of tailings containing high levels of mercury and arsenic pollutants into the ocean between 1994 and 2002.

The 20-month trial began after local villagers filed a US$543 million lawsuit against Newmont in August 2004, contending that tailings from its gold mine had caused serious illnesses and the ruin of their fishing income.

Initial assessment by the World Health Organization and the Indonesian Ministry of Environment found the water to be unpolluted, but a subsequent ministry study found that arsenic levels were 100 times higher at the wastedumping site than in other parts of the bay. The district court that cleared the case in December 2007 claimed that it could not prove the company polluted the bay, and the mine waste did not breach local or international safety standards.

More recently, in August 2008, the chief of a Papuan tribe in Mimika Barat Jauh, Kaimana district, West Papua, urged the government to take action against US mining company Freeport, and demanded compensation for contaminated rivers from mine tailings. The chief claimed that as a result of Freeport's mining operations, local village communities were facing water shortages, and that crocodiles and tortoises of Teluk Etna were on the brink of extinction.

In 2006, a toxic tailings spill at an Australian-run mine on Rapu-Rapu Island, Philippines, spurred mining opponents, including the Catholic Church, nationalistic politicians, and NGOs to demand a total ban on mining activities in the country. An independent commission report urged the company to stop all mining in the area, revoked its mining rights, and forced the company to pay damages. Philippine President Gloria Arroyo declined the commission's request to ban all mining activities, but promised to carefully study the report's recommendations, including a review of the 1995 Mining Act's provisions for foreign participation and management.

DEFORESTATION

Mining is one of the major causes of deforestation and forest degradation, as commercially valuable minerals are often found in the ground beneath forests. Large-scale, open-pit mining operations can result in significant deforestation through forest clearing in order to access mineral deposits and to open remote forest areas for miners. Infrastructure built for transient mine workers—roads, tunnels, and dams—also has an impact.

Large quantities of timber are often used as supports in mine shafts and in tunnels, in the case of underground mines, and as fuel for operating mines. This link is a major concern for India, where forest lands have been increasingly designated as mines in the last decade. Between 1980 and 1997, the Ministry of Environment and Forest (MoEF) only granted forest clearances for 317 mines, which destroyed a total of 34,527 hectares of forest area.

By contrast, from 1998 to 2005, the ministry cleared 881 mining projects in forest areas, destroying 60,476 hectares of forest area. As Chandra Bhushan of the Centre for Science and Environment in India observes, "This is an extremely worrying trend, since the mining industry has just begun expanding: with the new mining policy on the anvil, which is likely to give a significant boost to the industry, much more forest land is likely to be diverted for mining.

RIVER SAND MINING

Minerals and water resources are intimately linked in river sand mining. Extensive river sand mining substantially alters the physical, chemical, and biological environments of rivers. Mining and dredging activities, uncontrolled dumping of overburden, and chemical spills reduce water quality and poison aquatic life. Indiscriminate sand mining can make an area prone to flash floods. Sand mining has also been blamed for water shortage, as it deepens the riverbed and depletes groundwater.

Bank erosion and channel deepening from sand mining can also undermine nearby bridges and other engineering structures. With rapid urbanization and growth in the housing and infrastructure sector all over Asia, the demand for sand and gravel has grown significantly over the past few years. Most of this demand is being met by rampant, often illegal mining of riverbeds. In July **2008**, a professor observed the serious impact of sand quarrying in the Amaravathy

River in Chettipalayam, India: "Amaravathy used to provide drinking water for the entire Karur town but now it is not even able to supply water to the villages on its bank."

In the Palakkad district, the diminishing stock of sand and increasing environmental concern forced the state government to ban sand mining in the district. But illegal mining still takes place, and some have been selling river sand mixed with cheap sea sand, which causes structural failures in buildings. Kelani River, one of the major rivers in Sri Lanka, has also been badly affected by sand mining in recent years.

Overmining has caused problems such as salination of Colombo's drinking water and the collapse of the river bank. However, it is difficult to ban sand mining in the river because many people living nearby depend on it as a source of living. Sand mining has considerable implications for food security. It not only leads to water shortages affecting agriculture, but many farmers, seeing the high profits to be made, have abandoned their traditional ways of rice farming for it. Some in Bantul, Yogyakarta, have realized that sand mining is more profitable and less labour intensive than rice production.

Mining activities are threatening the forests of other Asian countries as well. In October 2008, the Indonesian Mining Advocacy Network (JATAM) and WALHI-Friends of the Earth Indonesia, protested BHP-Billiton's destruction of protected forest in Central Kalimantan for coal mining and in Gag Island, West Papua, for nickel mining. From 1999 to 2004, BHP-Billiton, along with other multinational mining companies, put pressure on the Indonesian government to amend Forestry Law 41/1999, which prohibits open-pit mining in protected forests.

Despite public protests across the country, a new law was issued in 2004 that permitted companies holding mining contracts issued before 1999 to operate in protected forests, including BHP-Billiton and PT Gag. In Central Kalimantan, BHPBilliton's coal mine affects 65,858 hectares of protected forest covering the upper reaches of the area's main rivers.

Forest degradation due to mining projects has other effects: it causes a decline in the natural environment's productivity, and it renders the local community more vulnerable to fatal floods, landslides, adverse climatic shocks, and other natural disasters.

Some governments in the region have begun taking action against the devastating impacts of sand mining. In November 2008, Chief Justice Sarath Silva voiced his concern over illegal sand mining in Sri Lanka. Together with Environment Minister Champika Ranawaka, he recently participated in a progrmme to promote awareness of the serious consequences of illegal sand mining.

The government of Andhra Pradesh, India, has taken measures to control illegal and excessive sand mining by seizing vehicles that have been ferrying sand. Sand mining became a transboundary issue in February 2007, when the previous usual supplier, Indonesia, abruptly banned its export of sand to Singapore, citing the impact of a recent Singaporean construction boom. Mari Pangestu, Indonesia's Trade Minister, banned the exports, saying it was necessary to protect the environment and maintain her country's maritime borders, threatened by rampant sand removal. Smugglers can ship as much as 10,000 cubic meters of sand from coastal areas in a day. Indonesia's Maritime Affairs Minister Freddy Numberi claimed that the ban was also aimed at pressing Singapore into signing a long-stalled extradition treaty. Singapore's Ministry of National Development expressed disappointment not only with the decision itself, but with Jakarta's rejection of Singapore's offer to help resolve Indonesia's environmental concerns. Not all sand mining has lasting effects if done responsibly; sand mining is nonchemical and no blasting is required.

However, there is still a need to increase awareness about the environmental impacts of increasing scarcity of sand and escalating construction costs. Research institutions in Asia are working to develop alternatives to river sand. The faculty of civil engineering of the Government College of Engineering in India has developed a technology to produce a cubical-shaped sand with excellent binding strength from hard granite.

THE QUANTITY AND QUALITY SUSTAINABLE AGRICULTURE

In simplest terms, sustainable agriculture is the production of food, fiber, or other plant or animal products using farming techniques that protect the environment, public health, human communities, and animal welfare. This form of agriculture enables us to produce healthful food without compromising future generations' ability to do the same.

Agriculture cannot be considered sustainable if the quantity and quality of its products are inadequate to sustain the healthy survival of all who are dependent. Sustainable agriculture must be assessed by dynamic not static criteria. Simple maintenance of the *status quo ante* cannot ensure food security for expanding populations. From reliable assessments it appears that more than 5 million pre-school children throughout the world die every year from maladies caused or aggravated by malnutrition. Many causes of the melancholy state of, and forecasts for, future food and agriculture are offered:

a. a selfish indisposition of the affluent, both among and within nations, to be more conservative in their consumption habits, to share essential resources more equitably;
b. politicians who, to retain power and influence, invest more in armaments than to sustain more efficient systems of food and agricultural production and distribution;
c. sects more intent on inflicting their particular ideologies on the rest of mankind than in helping to alleviate poverty, social and economic inequities;

d. self-proclaimed authorities who advocate a specific set of production principles or a production system sustainable for the whole world.

SUSTAINABLE AGRICULTURE IN THE UNITED KINGDOM

A report submitted to the British Minister for Environment, Food and Rural Affairs by a policy commission on the future of farming and food in Britain begins with a 'Vision' of desirable objectives and conditions for food and agriculture.

a. British farming and food sectors must be profitable, able to compete internationally, be good stewards of the environment, and provide a healthy, nutritious diet accessible to consumers in all income groups.
b. Production, processing and marketing must be holistically integrated, with more effective inter-communication among them.
c. The government is responsible for food safety, nutrition policy, animal welfare and environmental protection.
d. Public policy must recognise the interdependence of town and country.

The Report then compares its idealistic 'Vision' with the 'Present Situation' and states:

a. Farming and food industries [in Britain] are on a path that cannot be sustained over the long term.
b. Because of subsidies and price supports imposed by the EU Common Agricultural Policy, British and EU consumers pay more for food than the rest of the world, inflicting particular hardship on poor consumers.
c. The rural environment has been damaged by years of intensive agricultural production; while some farmers have adopted environmentally friendly practices, many cannot afford to do so.
d. Though the 'Vision' emphasises more holistic integration, significant disconnection exists among farm producers, food processors and retailers; unsatisfactory inter-sector relations and communications are detrimental to overall sustainable efficiency.
e. British consumers are uneasy about the wholesomeness and safety of the foods they purchase.
f. many health problems among Britons result from people eating too much of the wrong kinds of foods.
g. farming has declined in importance as an economic activity, farmers' share of retail prices having fallen. The total food and agribusiness industry contributes about 8 per cent, farming contributes less than 1 per cent to GDP.
h. farming is a highly competitive activity; farmers must develop more sensitive communications with consumers.
i. over 90 per cent of foods are purchased from supermarkets, allowing them undue influence over farmers and consumers.
j. the government must accept responsibility to support more

economically efficient, sustainable farming and food systems; the government's role is to facilitate essential change, not to finance stagnation and protection of the status quo.

All of the issues raised in the Vision and Present Situation are discussed in considerable detail throughout the report.

SUSTAINABLE AGRICULTURE INITIATIVE

A few years ago the SAI was launched by the Unilever and Nestlé corporations, since when some 20 food processing and related agribusiness companies have become participants. Though there is sharing of general information, each participating company pursues its sustainable agriculture research and development independently with particular attention to raw materials essential to its brand products.

The quality of every processed food is critically dependent on the biochemical and biophysical properties of the raw materials from which it is processed. The properties of each raw material result from properties genetically inherited, degree of maturity at harvest, post-harvest conditions of handling, storage, transportation and the processing technologies applied in food factories. The SAI programme seeks to establish and maintain close cooperation between farm producers and processors: to improve and, where necessary, modify farming practices to ensure delivery to processors of raw materials of defined acceptable properties. SAI encourages farming practices conservative of critical resources, protective of land, water, the environment, safety and wholesomeness of all raw materials. Listed among SAI objectives are:

a. To reduce risk of harmful residues such as pesticides, hormones and obnoxious contaminants;
b. To ensure safe on-farm working conditions;
c. To improve and expand farmers' knowledge and skills;
d. To preserve soil fertility, minimise erosion, encourage minimum tillage;
e. To promote farming systems that minimise waste and avoid pollution of surrounding environments;
f. To reduce use of chemical pesticides by adoption of biological and biopesticide control;
g. To use chemical fertilisers conservatively and effectively;
h. To conserve and protect scarce water resources;
i. To minimise energy consumed from non-renewable resources;
j. To explore practical economic alternatives to fossil fuels;
k. To protect biodiversity and wildlife habitats, the health and well-being of all farm livestock;
l. To employ environmentally benign production systems;

m. To raise farm productivity and profitability, to increase crop yields and reduce in-field and post-harvest loss;
n. To devise efficient farming systems based on systematic analysis and simulation modelling of all existing and potential biological, technological, social and economic parameters.

The primary objective of the SAI is to ensure long-term availability of agricultural raw materials essential to their industrial processes, materials that provide more than 2/3 of all raw materials used by these industries. SAI participants' dedication to the above objectives is highly commendable, motivated by the need to restrict extensive degradation of arable land; recognising the competition for land from urban and industrial spread, and the desirability of protecting biodiversity, nature reserves and wild life habitats.

SAI seeks to respond to the worldwide shortage of fresh water of which agriculture's consumption is close to 80 per cent of total world supply. SAI recognises growing competition from industries and urban communities, pollution of ground and surface waters by agricultural chemicals, municipal and industrial effluents, acid rain caused by industrial toxic emissions.

Guided by overall 'sustainable' objectives, SAI participants' R and D focuses primarily on agricultural crops and livestock of specific importance to their companies' processes. Unilever's concentration is on tea, palm oil, various vegetables and fisheries; Nestlé gives precedence to milk, coffee, sugar and cocoa; Danone to cereals and fodder crops. Environmental protection and resource conservation are also reflected in SAI participants' factory operations, where they pursue energy and water conservation, reduction and recycling of packaging and waste, elimination of refrigerants and other substances potentially harmful to atmospheric and environmental protection and conservation.

Though much maligned by 'anti-globalisation' activists, many transnational food and agribusiness corporations act more responsibly in conserving critical resources, protecting biodiversity, fragile ecologies and ecosystems, in improving agricultural practices and rural economies than can be said of some governments whose subsidies to farmers, damage to ecosystems and natural habitats in pursuit of oil and timber are destructive of ecologies and biodiversity. Governments and development agencies could benefit by following the SAI pursuit of economic, holistic integration of all logistical activities and processes that constitute a benign integrated food system that begins with sustainable on-farm production and ends with delivery of safe, nutritious foods to consumers.

INTERNATIONAL AGRICULTURAL RESEARCH IN SUSTAINABLE AGRICULTURAL SYSTEMS

A history of the foundation and early years of the Consultative Group on International Agricultural Research (CGIAR) was published by the World Bank. Other publications provide details of the growth and activities of the CGIAR and the family of International Agricultural Research Centres it supports. Annual Reports of the CGIAR and individual IARCs are available from the CGIAR Secretariat at the World Bank in Washington.

All of the CGIAR research centres protest their dedicated commitment to sustainable agricultural systems and practices. But it is not always clear whether the primary purpose is for each centre to be permanently sustained or to ensure that national agricultural organisations, the developing country farmers and the populations they serve are to be sustained. The CGIAR had its genesis in agricultural research sponsored by the Rockefeller and Ford Foundations. In 1943, Rockefeller began a bean and cereal improvement programme in Mexico, which eventually brought forth semi-dwarf high-yielding wheat phenotypes and, in 1963, metamorphosed into the International Wheat and Maize Improvement Centre (CIMMYT). In 1960 the two Foundations established the International Rice Research Institute (IRRI) in the Philippines, and later two centres to study tropical agriculture, one in Nigeria, and a second in Colombia. The CGIAR began in 1971 as an informal consortium of donor and development agencies with no legal constitution. From the start the World Bank has provided a supporting secretariat. Scientific assessments and recommendations come from the Technical Advisory Committee (TAC), a changing international group of experienced scientists with a secretariat housed at FAO in Rome.

In 1971 eleven members pledged $15 million to support the four existing IARCs. By 1976 membership had grown to 26 donors, who contributed $63 million to the four original and seven new IARCs. By 1992, CGIAR has expanded to a membership of 40 with a budget close to $300 million to maintain a family network of 16 IARCs and related programmes. The *CGIAR Report* for 2002 lists 62 donor members, including the three sponsors: World Bank, FAO and UNDP, and states there are now 16 IARCs that employ more than 8500 scientists and support staff. (A private communication indicates that in 2004 there were 15 centres and programmes, supported by over $400 million from 63 CGIAR members). The *2002 Report* refers to five priorities declared at the 2002 Johannesburg Conference: water, energy, health, agriculture, and biodiversity. Following a brief description of activities in each of the IARCs, the report describes Challenge Programmes in which several IARCS cooperate with emphasis on a. water and food, designed to increase productivity of water for food and livelihoods; b. bio-fortified crops for improved human nutrition: breeding crops with higher than normal contents of such micro-nutrients as vitamin A, iron and zinc. The report also describes a Global Conservation Trust, jointly sponsored by the CGIAR and FAO, to ensure diversity among the most important food crops through an endowment to provide funds sufficient to maintain crop genetic resources held in gene banks around the world. It mentions a report published by the International Peace Research Institute in Oslo that states that between 1989 and 1997 most of the major military and civil conflicts around the world occurred in regions heavily dependent on agriculture.

A working group convened at FAO in 1989 by the CGIAR Technical Advisory Committee, with representatives from IARCs and several donor agencies, concluded that, "whereas agricultural production and farming systems have been well served by the CGIAR family, post-production systems, including breeding for functional properties essential to food processors, have been largely ignored."

ORIGINAL PURPOSES AND PROGRESS

At the outset CGIAR and TAC agreed that the IARCs' research would be devoted to improvement of crops which provide most of the food for developing nations: crops such as rice, wheat, maize, sorghum, several grain legumes, cassava, potatoes and other root crops. Later IARCs were created for research on livestock health and husbandry, fisheries, forestry, agro-forestry, bananas, and a centre for collection and classification of plant genetic resources and the conservation of biodiversity. Each IARC responsible for one or more food crops maintains an extensive germ-plasm bank of broad genetic diversity. National agricultural research and development programmes can select and receive genotypes and phenotypes of crop species suited to their needs and ecologies. Each centre provides training in the disciplines and activities related to its research programme. The CGIAR and its IARCs have assembled the world's largest, most diverse collection of plant genetic resources, amounting to some 750,000 classified accessions. The early advances at IRRI and CIMMYT, much publicised as the 'Green Revolution', resulted from Mendelian plant breeding which shifted synthesised biomass from vegetative plant parts into the edible seeds. Plant breeders also developed and propagated rice types that matured more rapidly than farmers' traditional varieties. Rice farmers were thus able to plant and harvest two or three rice crops on the same land every year. During the 1960s and 1970s, rice production across Asia more than doubled; India increased its wheat harvest from 11 million tonnes in 1961 to 70mt in 2002.

Much early criticism complained that those who benefited most from higher yields were large-scale not small-scale producers with limited resources. The world's history of technological progress illustrates that innovative and novel technologies are first adapted and adopted by individuals and industries with access to resources sufficient to tolerate the inevitable risk of investing in unfamiliar practices. Eventually, the higher yielding crops and related production technologies were accepted by a majority of Asian rice farmers.

FARMING SYSTEMS RESEARCH

During the 1970s IRRI scientists laid the foundation for modern concepts of farming systems research, in a programme that began modestly with a few Filipino farmers but gradually spread to include many hundreds of rice farmers in 12 Asian nations. Western agricultural research traditionally began and continued on an experimental farm until promising results were eventually transmitted to farmers via extension services. In contrast, the IRRI farming systems research methods begin by assessing opportunities for, constraints and risks to increasing production by adaptation of new crop types and production technologies. The methodology starts with an analytical assessment of the biological, technological, physical, financial, economic and ecological resources existent among different farming communities; how and with what resources could more productive technologies be introduced with tolerable risk. The IRRI farming systems methodology has contributed significantly to increased crop production while conserving farmers' ecological and economic resources.

ESTIMATED BENEFITS

Evenson and Gollin [2003] estimate results from what are variously designated 'high-yielding varieties' [HYV] and 'modern varieties' [MV] of food crops grown in developing countries. They compare changes in yield and total production during the 'Early Green Revolution (1961–80)' and the 'Late Green Revolution (1981–2000)' and the calculated influence of MVs mostly developed in CGIAR/IARCs. From 1961 to 1980 MVs accounted for 21 per cent of increased yield and 17 per cent of production increase; from 1981 to 2000, for 50 per cent of yield growth and 40 per cent of production increase. Most notable increases attributable to MVs were in Asia, the Middle East and Latin America. Little benefit from MVs was realised in sub-Saharan Africa where most increases resulted from expanding land under cultivation. Using a mathematical simulation model, the authors estimate that, without MVs in the period 1960 to 2000, food crop yields in developing countries would have fallen. The authors also claim that without MVs food prices would now be significantly higher throughout the world.

SUBCONTRACTING TO NATIONAL PROGRAMMES

An initiative that started in the International Centre for Research on Potatoes (CIP), and later pursued by IRRI, subcontracts programmes and projects to national agricultural research institutions in developing countries. The IARCs have trained thousands of scientists from poorer nations many of whom return to their native lands without funds and resources adequate to apply the knowledge and skills they have acquired. Subcontracting IARC programmes and projects to national research bodies serves a. to reduce operational costs, as developing country scientists cost less than their international counterparts; b. to increase resources and strengthen capabilities in national programmes. For agriculture to satisfy the needs of Africans, Asians and Latin Americans, resources and responsibilities should be systematically transferred from CGIAR/IARCs to national programmes and institutions.

6

Sustainable Forest Management

Criteria, indicators and standards for sustainable forest management and for SFM are important market based water demand management tools that foresters can use for improving the water application and water productivity elements in WC&DM. The organised industry has adopted the Forest Industry Environmental Guidelines for Conservation of Commercial Plantations in South Africa, which include water management measures.

WATER MANAGEMENT PLANS

The National Water Policy requires sectoral management plans. These are to incorporate into the WC&DM Strategies. This section outlines the requirements for the water management plans in the forestry sector.

Water management plans may be formulated at three scales. The first would be for the forest management unit (FMU), *i.e.* as defined in the National Forest Act as "an area of land on all or part of which there is forest and which is managed as an integrated unit".

The second would be for a Water User Association comprising of several owners of forest management units within a Water Management Area. The third would be for the sector as a whole. What follows are requirements for plans at the level of FMU. Plans at the other levels can be built from these guidelines. These requirements are built upon the "Environmental Guidelines for Conservation Management of Commercial Plantations in South Africa" (second edition, in press), and the minimum standards for forest certification according to the Forestry Stewardship Council (FSC) as applied in South Africa. These have been enhanced to take account of the requirements of national water policy. South Africa has adopted principles of sustainable forest management (SFM), which are in harmony with international principles, and which are incorporated in the National Forests Act. These will guide the development and regulation of criteria, indicators, and standards of SFM in South Africa.

Forest management requires as standard practice the development of detailed forest management plans for each forest management unit. These plans deal in detail with the full cycle of forest production, from seedling through final harvest, as well as overall environmental management of the FMU. The Environmental Guidelines for Conservation Management of Commercial Plantations in South Africa stipulate an inclusive set of environmental guidelines that should be followed in forest management planning. The FSC also sets standards to which the plans and their execution must comply if the FMU is to be certified as sustainably managed. The incentive for certification is strong, since a grower wishing to sell forest produce to a processor who exports, needs such a certificate. Thus, water management planning on the FMU must be an integral part of the sustainable forest management plan for a FMU. Whether or not the water management plan is adequate must be judged by the overall adequacy of the forest management plant itself. Generally, the certification of a forest can be taken as an adequate test of the water management plan if clear standards are available, since the first requirement of certification is compliance with all relevant laws.

WATER AUDITS IN THE WATER MANAGEMENT PLAN

The first section of the water management plan must consist of a documented audit of the water resource of the FMU. This audit must be based upon the best available information, such as the recorder water balance (precipitation and run-off), flow regime, and water quality regime of the catchments or parts of catchment included in the FMU, or the best predictions from appropriate hydrological models. The audit must also include an assessment of the relevant elements of the current management plan for the FMU.

The documented audit must contain at least the following:

- A quantification of the water balance of the FMU under natural conditions, *i.e.* as it would have been without afforestation, under natural vegetation, during the period since afforestation, together with the flow-duration curve after afforestation.
- A quantification of the water balance and flow-duration curve after afforestation.
- The same for the water quality regime, and an assessment of how this departs from the resource-quality objectives for the river reach fed by the FMU, as set out in the DWAF Resource-Directed Measures.
- From all this, an assessment of current water use, allocation efficiency, application efficiency, and productive efficiency.
- An assessment of the compliance of the water management elements of the forest management plan to accepted minimum standards.

Management Goals and Objectives

The water management plan must contain a statement of overall goal and of contributory objectives, in terms of mean annual run-off, the flow-duration curve, and water quality objectives, as well as management standards. These goals must take account of normal climatic variation and predicted climate change.

Elements of Water Management

The principles that apply here come from the National Forest Act, the National Water Act, and the National Environmental Management Act, and are summarised in the Environmental Guidelines for Conservation Management of Commercial Plantations in South Africa. The South African Forestry Industry is committed to Integrated Environmental Management to ensure that:

- Development takes place in the most economic and environmentally acceptable way;
- Resources are managed in a manner which will ensure the sustainability of the forestry enterprise; and
- The people on which the Industry depends may work in safety and live under conditions of acceptable quality.

Principles of Sustainable Forest Management (SFM)

The National Forests Act of 1998 contains the following six principles to guide SFM:

- Maintain the resource base
- Maintain the sustained yield of goods and services
- Maintain biodiversity
- Wise use of water
- Optimise the socio-economic impacts of forestry
- Develop supportive institutional frameworks and policy

The Environmental Guidelines for Conservation Management of Commercial Plantations in South Africa recognises the following as some of the values that could be associated with the above Principles of SFM:

Soil

- Soil quality
- Soil quantity
- Nutrient status

Water

- Water quality
- Water quantity
- Wetland habitat

Biodiversity

- Species diversity

- Ecosystem function
- Important species

Forest Health

- Protection against pests
- Protection against diseases
- Protection against fire

These form the framework for detailed guidelines on several aspects of environmental management on the FMU, including the following:

- fire protection: adequate and appropriate fire management strategies are imperative to ensure that the forest resource, together with its ecological and other associated values, are protected
- Site/species matching: to ensure optimum yields, quality and economic benefits, the requirements of the species should be matched to the characteristics of the site
- Avoiding monoculture landscapes: monocultures pose a risk in terms of disease and fire, which are heightened when single clones are utilised; If monocultures are unavoidable, a variety of age classes is a suggested improvement for reduction of risk and to improve the general aesthetics.
- Weed potential
- Water use efficiency: efficient species, clones and management regimes should be favoured
- Wetland protection.
- harvesting system: The environment should determine the type of harvesting system to be used. In choosing the most appropriate system, terrain, weather, extraction routes, timetables and yarding sites should be considered: detailed guidelines for environmental protection are applicable

roads: the exposed surfaces of roads are the main source of sedimentation in the plantation environment; by restricting this disturbance as much as possible, this impact and the loss of productive land can be reduced: detailed guidelines for environmental protection are applicable.

INSTITUTIONAL SUSTAINABILITY: A CONCEPTUAL FRAMEWORK

To explain institutional sustainability, analysts often rely on ad hoc arguments. The factors they stress include national level commitment, in particular a willingness to provide for recurrent costs; an appropriate mix of public and private sector institutions; within the public sector, an appropriate delegation of tasks to the various levels of government; a sensitivity to issues of ecological degradation and long-term agricultural productivity; and adoption of management systems that promote capacity building. Each of these special explanations is plausible, but they do not necessarily fit together or complement each other. We therefore propose a new overall framework for understanding sustainability, one that consolidates many of the partial analyses of why institutions do or do not maintain themselves over time. Our formalization draws freely from systems theory, contingency theory, and political economy. Hence we

This chapter grows out of research sponsored by the U.S. Agency for International Development's (USAID) Office of Agriculture and Rural Development in the Asia/Near East Bureau to address the problem of development sustainability. The bureau's effort grew out of earlier work on the topic undertaken by the University of Maryland's International Development Management Centre and the Department of Agriculture's Development Program Management Centre (DPMC) for another office in USAID, the Office of Institutional and Rural Development in the Science and Technology Bureau. have dubbed it SCOPE, the acronym standing for its three constituent bodies of thought. In this chapter we lay out the SCOPE framework.

We build from the idea that institutions are systems. There are several reasons for starting on this foundation. First, systems theory is general enough to cover sustainability phenomena in institutions of all kinds-from formal collectivities to informal codes of conduct.

Second, systems theory is concerned simultaneously with internal processes and the relationship between the system and its environment. It thus forces us to look at a wide variety of social, economic, political, technical, and other factors that affect sustainability. Finally, systems theory emphasizes interrelationships among disparate elements, and so enables us to merge the agro-environmental, economic, and managerial aspects of sustainability.

Systems theory, however, provides little guidance about how to portray internal system processes or external system relationships. Contingency theory fills in this gap. It gives insight into how institutions can best attain congruence with the forces around them. Another reason we employ contingency theory is to avoid the tendency, widespread in the literature on institutional sustainability, to recommend all-embracing, generic actions. We explicitly reject cure-all solutions. In analyzing the long-term maintenance of institutions, it is also critical that key dimensions of the external environment be characterized. Here is where political economy helps. This school of thought focuses on the exchange of resources, and provides penetrating means to characterize the structure and tasks of institutions. By assisting in identifying the circumstances under which the different kinds of resources will be produced and traded, it also amplifies the contingency approach.

We do not claim to have evolved a rigorous theory of institutional sustainability, one that emerges directly from a set of fundamental propositions.

Rather, SCOPE is a heuristic device, intended to provide a vocabulary and perspective that incorporates disparate ideas about sustainability.

This does not mean our conceptual framework is atheoretical. The precepts we pull together are rooted in three well-developed schools of thought. And despite their different origins, they are mutually reinforcing, not just a patchwork of ideas.

S IS FOR SYSTEMS

At the heart of systems theory is the idea that many natural and social phenomena can be treated as a "system" — defined broadly as a set of units or elements that interact with one another to convert some types of inputs into some types of outputs. Analysts have found systems every where, in something as simple as a biological cell to something as complex as a nation state. In an interdisciplinary, comparative field like international development, the idea of a unified mode of analysis is appealing.

And indeed, systems theory has long been applied to development topics. But practical use of the theory has been hampered by its abstractness. In their search for broad and general terms with which to analyze systems of all shapes, sizes and levels of complexity, disciples of this school have created a dense and idiosyncratic vocabulary. We endeavor to avoid the more abstruse of these formulations, and to remain grounded in developing country experience. From the viewpoint of social science, systems theory has been most usefully employed in analyzing formal institutions or organizations, such as government agencies, business firms, and universities.

According to Hage and Aiken, what such collectivities have in common is that:

1. They are created and continue to exist to achieve a common purpose greater than that which could be achieved by individual action;
2. They develop specialized subsystems to subdivide goals and increase efficiency;
3. They use patterns of authority and communication relationships to link and integrate specialized subsystem activities;
4. They depend upon transactions with external clientele and suppliers to provide inputs or absorb outputs necessary to maintain and develop the system.

Sustainability, of course, is not just a problem for formal organizations. It also concerns looser, less clearly defined institutional arrangements-a regional network of farm markets, for instance.

One of the advantages of systems theory is its ability to handle both this kind of institution and the more formalized variety. In this chapter, we will discuss how the SCOPE framework can be used to explicate sustainability across the range of institutions.

In looking at human organizations as systems, several key definitions need to be kept in mind:

1. Inputs (resources) — the raw materials, money, people, information, and knowledge that an organization obtains from its environment and that contribute to the creation of its outputs;
2. Outputs — products, services, programs, and ideas that are the outcomes of organizational action. Organizations transfer their major outputs to the environment, and use others internally;
3. Technology — methods and processes (mental, physical, mechanical) for transforming resources into outputs;
4. Environment — comprised of the *task environment* (external organizations and conditions that affect an organization's main operations and technologies, such as clients, competitors, markets) and the *general environment* (institutions and conditions that have long range impact on the organization and its tasks, for instance the economy, the political system, the state of scientific knowledge, or the national culture within which the system operates);
5. Purposes — strategies, goals, objectives, plans of the organization's dominant decision makers;
6. Behavior and processes — prevailing patterns of behaviour, relationships between groups and individuals;
7. Culture — shared norms, beliefs, and values of organizational life;
8. Structure — enduring relationships among individuals, groups, and larger units.

An important precept of systems theory is the ordering of systems according to hierarchy. All but the simplest systems are made up of smaller systems, which in turn are made up of even smaller ones, and so forth. Thus a nation state embraces communities, which contains cliques and families, which are comprised of individual human beings.

One implication is that any system's environment is also a system in its own right. These external systems influence the flow of inputs to institutions, affect the reception of outputs, and may stimulate changes in internal operations.

Another implication is that the constituent parts of an institution can be seen as sub-systems, driven by their own need to acquire and process resources. Institutions are influenced by these sub-systems as well as their environment. Bureaus, offices, divisions, and other units may contribute to the larger organization's operations, resist them, or change them from within.

CO Is for Contingency

Contingency theory has been an important strand of administrative science since it was first articulated in the mid-1960s. Its major premise is that the optimal structure and management style for an organization are *contingent* on uncertain, exogenous conditions. Contingency theory thus shares with systems analysis a concern for environment.

Every human aggregation or pattern of behaviour has to be seen in relation to the complex of outside forces that threaten or promote its survival and expansion. This school of thought within the management sciences disputes the idea that any particular model of organization can truly be considered "best." The question posed is, best for what?

Contingency theory's basic postulate stands in contrast to the assumption, often made implicitly in social and economic theory, that there exists a single, optimal way to constitute formal collectivities. Even so astute an observer as Weber, seems to have held this position, for in his discussion of bureaucracy, he maintained that its superiority would enable it to spread everywhere in modern society. Bureaucracies have indeed become ubiquitous. It does not necessarily follow, however, that a bureaucratic model is advisable for all types of collective action.

To the contrary, when they began to look at real organizations in depth, researchers were surprised to uncover a wider range of successful adaptations than they had initially assumed possible. The Weberian model was found, not surprisingly, to be generally efficient in mass-producing standardized goods or services.

Marked by hierarchy, precisely defined roles, and functional specialization, this type of organization performed well in situations that were competitive (so it was important to produce at low cost per unit of output), that were characterized by repetitive tasks (making it easier to divide up labor), and that used mature, stable technologies (meaning innovation was relatively unimportant).

But the same model did poorly under the opposite sort of conditions, in which cost was a less important consideration, in which jobs were not repetitive, or in which new technology had to be rapidly adopted. It was discovered that in this domain, less bureaucratic modes of organization were better suited.

The basic propositions that emerge from this line of analysis are the following: (1) No universal organizational principles are right for all situations; (2) The success of an organization depends on it having an adequate "fit" with its environment; (3) To attain this "fit," an organization must possess appropriate structures, strategies, cultures, and so on.

Since the mix of contingencies differs from setting to setting, what works well for one enterprise may be unsuited for another. The important point is that the organization has to be able to adapt to its environment — or find a favorable one in which to operate.

Given the diversity of cultural, political, and administrative conditions prevailing in less developed countries, contingency theory figures prominently in current thinking about development management. For example, when Honadle and VanSant observe that special project management units have proven effective in managing civil works projects or those social service projects with a known technology (such as an immunization campaign), but have been an ineffective way to deliver services, they are making a contingency argument.

So is Paul, when he makes the point that successful organizations are those that figure out which tasks need to be centralized and which decentralized. In both instances, the logic follows the same path: the most effective structure of an organization is contingent on the particular factors that confront it.

P Is for Political, E Is for Economy

The term political economy has come to refer to two rather distinct schools of thought. One usage stresses that politics and economics are inextricably linked and must be examined in totality. Political economy in this sense descends from the writings of classical political economists, starting with Adam Smith and ending with Karl Marx, all of whom emphasized that the creation and distribution of wealth were not merely private affairs, but were important matters of state, too. Their intellectual successors today often take radical positions, or at least ones that are critical of contemporary attempts by governments to regulate market forces for the common good.

Political economy in its other sense applies neoclassical microeconomics to political behaviour. Relying on methodological individualism, this tradition focuses on "political man" (and woman) who is the exact analogue of "economic man." It sees people in public life as the same purposeful, utility maximizers that they are in private. Political man runs for office, votes, joins political parties, and supports interest groups to advance his own well-being, as he defines it.

The intellectual wellspring of this tradition is Schumpeter, whose basic insight has been amplified by Downs and others. Unlike the other strand of political economy, this one (which also goes by the rubric "public choice") is more often associated with conservatism. Its proponents tend to take a jaundiced view of human nature and to be skeptical about the prospects for social and economic reform through collective action. Despite the ideological contrast between these two varieties of political economy, they are not philosophically incompatible.

In particular, they share the assumption that political behaviour is dominated by economic considerations and, conversely, that the production and consumption of goods and services are shaped by legal and governmental practices. An important implication is that political resources can be parlayed into economic influence, and economic resources into political power. (These ideas, of course, are consistent with systems theory.

Using its language, the polity and economy are systems that influence each other.) In this line of analysis we find many insights into how institutions interact with their environment, particularly with their constituencies or "stakeholders" (more on this topic follows).

Both varieties of political economy also maintain that economic institutions (such as markets or firms) and political institutions (such as legislatures or bureaucracies) fulfill parallel social functions. Most prominent of these functions are: (1) to aggregate individual choices or preferences into social choices or preferences; and (2) to reduce the costs of interactions or transactions among elements in society. Political economy has rich theoretical tools for evaluating and interpreting how efficiently institutions carry out these tasks.

It is not surprising (especially in light of contingency theory) that political and economic arrangements vary widely in their ability to sum up preferences or hold down transactions costs. The case of rural influence on national economic policy is an example. One of the challenges facing small to medium level farmers throughout the Third World is to organize themselves as an effective pressure group, the way they have in most Western countries.

All other things equal, farmers have the easiest time pushing their claims in nations with competitive political institutions (such as India), or nations that have relatively egalitarian economic systems, particularly regarding the distribution of property (such as South Korea). Political economy is the best body of theory to explain this facet of institutional sustainability.

EXTERNAL SUPPLY AND DEMAND

Since systems theory considers the environment critical, we begin by characterizing environment-system interactions. The main external influence on institutions takes place through factor and product markets. This can occur either directly or indirectly. Let us discuss *direct influence* first. This entails no intervening steps or factors. The environment has an immediate effect on inputs to the system, its outputs, or its structures and processes. Particularly important are shifts in the amount or valuation of a system's "raw materials" and its "finished products." An example is, where a ministry of education sets new entrance requirements for post-secondary students, thereby forcing a university to modify its mix of course offerings and, ultimately, of faculty. Internal practices have changed in direct response to external demand. We stress that demand is not a narrow economic concept. What a system's inputs or outputs are "worth" is more than an objective measure of energy produced or marginal costs. It is a subjective measure of utility. As Porter puts it when talking about business firms seeking a sustainable competitive advantage over their competitors, "value is what buyers are willing to pay for." In the field of agricultural development, the "buyers" are not always immediately obvious, as they are in private industry. But they may be even more important.

This is especially true when the "buyers" are political elites. Their attitudes explain how a system may survive while generating few economic goods or services. For example, the seeming contradiction of parastatals in Africa: These are inefficient systems in conventional terms — the ratio of tangible outputs to inputs is low.

Their weaknesses are well-known, and, indeed, have often been the target of donor initiated reforms. Yet parastatals persist. Why? Partly because African elites prize political control over key economic sectors. A similar example is government management of agricultural production, which in many developing countries keeps producer prices at unremunerative levels to provide inexpensive food to the politically salient urban areas. In this instance, elites rate political calm in the cities above economic health in the countryside.

One of the most important influences on value is whether a system produces "public" or "private" goods. Political economy defines private goods as items that can be used exclusively on an individual basis. Public goods, by contrast, are nonexcludable or can be jointly consumed. Nonexcludability means it is difficult to prevent other people from using the good once it has been provided to one person.

Common rangeland, which is available for all pastoralists to graze their cattle, is an example of a nonexcludable good. Joint consumption signifies that many people can enjoy the good without diminishing each other's enjoyment. Rural roads and flood control dams (but not common rangeland!) are examples of goods that, within broad limits, can be jointly consumed. The special properties of public goods are attracting new attention in development management circles.

There are few pure public goods, but many goods that have elements of nonexcludability or jointness of consumption. Even primarily private goods usually have some public dimension to them, or what economists call an "externality." Often these externalities put a hidden cost on society at large, for example, as occurs with charcoal in the Sahel.

This fuel is bought and sold as a private good, but its manufacture has public consequences due to the destruction of forests and ensuing desertification. Because private markets cannot make a complete reckoning of the costs of public goods, the price mechanism gets thrown off, and supply and demand schedules fail to meet where they should. This problem is the main rationale for giving an economic role to nonmarket institutions, since alternative arrangements will not meet the latent demand for public goods or charge prices that fully account for externalities.

By the same token, a system that supplies public goods can have a hard time generating support for its output. The reason is the individual behavioral incentives that result, incentives which tend to encourage "free-riding" or shirking. Planners often underestimate the problem.

An integrated rural development project in Jamaica, for example, tried to protect soil from erosion (primarily a public good) by paying hillside farmers to build terraces and ditches (primarily a private good). The farmers perceived little individual benefit from soil conservation itself, especially in relation to maintenance costs, and after project subsidies were removed they allowed the earthworks to fall into disrepair.

On the other hand, community solidarity or other cooperative norms can, on occasion, overcome the social inertia that accompanies the provision of public goods. This phenomenon is gaining increasing attention in the management of business firms, where conscious efforts are now widely made to promote so-called corporate cultures that encourage people in the organization to pull together for the good of all.

Hirschman notes that a similar, though spontaneous, "social energy" unites many grass-roots organizations in the Third World. In our terminology, cooperative action is due to value's many forms, social and cultural as well as political and economic. Adherence to certain norms, such as consensual decision making or centralized authority patterns, can be prized in its own right. A given system will be sustained only if it produces outputs people really want.

What about *indirect influence* from the environment to the system?

Unlike the value considerations just discussed, these are external conditions, incentives, or policies that affect a system's internal processes via some intervening step or factor(s). Primary education is an example. It has an indirect influence on a farming system's cost structure. Better schooling will not affect the price or labor in the first round, but in the second round it might improve the productivity and adaptability of farm workers — although could be offset by new attitudes towards manual jobs that would tighten up the market for field hands.

The point is that a far-off action, for example, a decision at the national level to provide universal primary education, could indirectly help determine the cost of inputs to a local agricultural system.

We classify three types of indirect environmental influence, each of which is particularly important for sustainability: *stability, flexibility,* and *artificiality.* Each of them is worth discussing in turn, to show how they affect what goes on inside an institution.

Stability

Stability (or the rate of change) in the environment is an important source of indirect influence on a system because it permits greater certainty in setting up internal processes. Institutions are constantly changing as relationships with outside elements shift. Small external changes may require no more than routine corrections elsewhere within the system, but major changes can trigger a system-wide series of adjustments that may prove too much to handle.

It is not surprising that institutions often stumble when they face even moderate environmental change. The more sudden and wrenching the change, the harder it is to cope with. One need only look at the disastrous organizational consequences of civil war in parts of Africa and Central America to see the importance of a minimum level of societal stability and predictability for institutions to maintain their vitality.

Flexibility

Flexibility (or permissiveness) in the environment is important because it conditions the possibilities for system change. For instance, if the environment is rigid and unforgiving, due to cultural norms that discourage experimentation, to try new patterns of internal organization may be out of the question, thwarting performance and threatening system vitality.

As with instability, the scarce resources that have to be used up coping with an inflexible, impermissive environment may mean that too few are left over to maintain system performance. Rigidities in the environment are frequently cited as a major hurdle for organizations in developing countries. Not surprisingly, many remedies for development problems entail a boost in flexibility — for example, calls for bureaucratic reorientation, administrative decentralization, or privatization.

Artificiality

Another important facet of a system's environment is its artificiality, or the extent to which resources from afar (*i.e.*, outside the immediate environment) are available. Artificial environments can be nurturing, as long as the high level of additional inputs is maintained. But they pose a hazard to institutions by encouraging overexpansion and external dependency.

A good example is a project management unit created to execute a donor-funded development project. Typically, these units' operating expenses are paid from abroad. Thus they have more and better equipment, more and better-paid staff, and can perform better in the short run than any host country organization with comparable responsibilities. But they often collapse once their foreign support is terminated. While all donor projects are artificial systems, there is a range of artificiality associated with them. On the low end is what might be called the seed grant approach, frequently used by private agencies such as the Ford and Rockefeller Foundations. Here the idea is to provide a few critical resources to induce the host country environment to provide the principal inputs. This has proven effective in building sustainable educational and research institutions in India, Mexico, and other countries.

At the high end of artificiality is the bypass approach that donors sometimes choose for particularly hostile country environments (for instance, Haiti or Zaire) with the intention of creating systems that can be isolated and controlled from the outside. These types of highly artificial systems are not seriously intended to survive without continued foreign support.

Most projects fall between these two extremes — an intervention to build feeder roads that uses local labor and simple technology would tend to be less artificial (especially if the host government is prepared to support maintenance costs), whereas an integrated rural development project with subsidized credit, infrastructure, and marketing services would tend to be more artificial.

Artificiality has an important macroeconomic dimension, too. In recent years international donors have been encouraging developing countries "to get prices right". Part of the reason is that distorted price signals create unreal environments, which in our terminology, promote artificial systems. Thus the push for devaluing national currencies, raising interest rates, and allowing prices to respond to market forces can be seen as based on an argument about the unsustainability of artificial environments.

Unrealistically high exchange rates mask the true cost of imports and promote excess consumption of foreign-made inputs. The resulting dislocations stunt growth and jeopardize the ability of all national institutions to continue purchasing the inputs they really do need.

PRINCIPLES OF SUSTAINABLE FOREST MANAGEMENT

ORGANIZATIONAL STRUCTURE OF FOREST

All forests are managed under the prescriptions of a *working plan*/scheme prepared on the basis of principles of sustainable forest management and recognized and innovative silvicultural practices. The authority as designated by the Ministry of Environment & Forests, Government of India, approves the *working plan* for this purpose. Generally no timber harvesting is done in any forest area without an approved *working plan*/scheme.

It is the duty of the owner/manager of the forest area to ensure the preparation of the *working plan*/scheme. While a detailed *working plan* is prepared for large areas such as forest division, working schemes are prepared for smaller areas for a specific purpose or areas like private, village, municipal, cantonment forests, etc. Even working schemes have all major elements of *working plan*; and these schemes also need the sanction of competent authority designated by the Ministry of Environment & Forests.

ORGANIZATION CONSISTS OF FORESTS

In the Ministry of Environment & Forests, the organization consists of Director General of Forests & Special Secretary to the Government of India, Additional Director General of Forests, Inspector General of Forests, Deputy Inspector General of Forests, and Assistant Inspector General of Forests. In the field, it consists of officers of Regional Offices headed by the Regional Chief Conservator of Forests.

In the States, there is no uniformity in the constitution of the organization. There are various levels of which Conservator of Forests Working Plan and

Working Plan Officer (DCF) are the key functionaries. Overall situation is as under:

- Policy/Supervisor Level – PCCF/APCCF/CCF;
- Field Supervisor Level –Chief Conservator/Conservator of Forests Working Plan;
- Field Function Level – Working Plan Officer (DCF/CF) assisted by ACF, RFO, and Foresters.

The number of *working plan circles* and *working plan divisions* depends upon the workload, *i.e.*, number of *territorial divisions* for which working plans are to be prepared/revised. *Working plan* is generally revised every 10 years; and on an average, one Conservator supervises the work of four *working plan divisions*.

The *working plan* of forests other than those under the control of Forest Department like municipal, cantonment, private, village, etc., forests can be prepared by Working Plan Officer on the request of the owner by owner themselves or through outside consultants. Micro-plan of jointly managed forests is prepared by forest staff of *territorial division* as per MOU and in consultation with the communities involved. The micro-plan so prepared is integrated with the *working plan* of that *territorial division*.

IDENTIFICATION OF FOREST AREA

The identification of the forest division whose normal *working plan* is to be prepared or revised is done by the Principal Chief Conservator of Forests in consultation with the senior officers of the Working Plan Organization. For other areas, this is mainly demand driven.

In JFM areas, it is on the basis of a MOU between the local community and the Forests Department. The task of preparation/revision of *working plan* is undertaken by WPO for the forest divisions within his jurisdiction as an essential function of the division. In addition, the PCCF or the competent authority may assign to him the task of preparing/revising *working plan* of other forest divisions to meet the target.

TENURE OF WPO AND OTHER SUPPORTING STAFF

For the preparation of the *working plan* of a *territorial division*, normally two years time is required. It may be less or more depending upon the volume of work and technical facilities available. The officers and the staff of *working plan division* should not be transferred during the preparation of *working plan.*

Head Quarter of Working Plan Division

Head quarter of *territorial division* is generally the head quarter of *working plan division* as well. It facilitates proper coordination and smooth flow of information/records between *territorial and working plan division.*

Status and Allowances

A *working plan division* is a functional charge, and the WPO has the status and power, unless otherwise stated in any particular respect, of a Divisional Forest Officer/Conservator of Forests as the case may be. A WPO is entitled to special pay admissible under Pay Rules applicable. Special pay/allowances should also be given to other supporting staff of the WPO.

BUDGET AND ACCOUNTS OF FORESTS

This is regulated by the rules and regulations of the State/UT Government. It is ensured that adequate budget provision is made in time. The proposed estimate is prepared by the Working Plan Conservator a year before the *working plan* revision is due to commence and copies sent to the DFO and the Territorial Conservator concerned also. The WPO is delegated the power of drawing and disbursing officer. Office management, maintenance of records, reporting of progress of work, and expenditure are according to the procedures laid down by the State/UT Government concerned.

Co-operation of Territorial Staff

It shall be mandatory for the DFO and the Circle Conservator to extend full cooperation and assistance to the WPO for fieldwork and provide logistical support, access to official records and other information so that the WPO can prepare good quality *working plan* expeditiously.

Office and Residences

Every State and UT Government should make efforts to provide necessary office and residential accommodation to all officials of Working Plan Organization both at head quarter and field levels as per their rules and norms in force applicable to their rank and status.

ELEMENTS OF THE STRATEGY RELATED TO RESEARCH

WC&DM has only recently been introduced into policy through the Act. There is still a great deal of research necessary to understand the requirements for its implementation. There is a need to better understand water use in the forest sector, water management options and the impact of external influences. In addition, consensus needs to be reached on the criteria to define concepts in the Act such as equity, ecosystem functioning and beneficial use. This baseline information is necessary to enable the forest sector to make informed decisions on WC&DM which promote sustainable use of water resources.

Considerable research has been undertaken to understand water use in the forest sector. This knowledge now needs to be framed within the context of WC&DM. A strong feeling has been expressed by stakeholders that there is a lack of good qualitative and quantitative, long-term information on water use in the forest sector. This includes the distribution of water use, the relationship between forestry water use and ecosystems, catchment specific information, impacts on groundwater, and the water use by different species in different locations.

There is no agreed method for defining or measuring water use in the forest sector and this is a significant obstacle which needs to be overcome. A key aspect may be an analysis of how a reduction in demand will impact on the performance of the sector. This will lead on to a consideration of water management options.

Current initiatives:

- The WRC has commissioned a three year project to the development of models for quantifying stream flow reduction caused by commercial afforestation in South Africa. This project is jointly run by the CSIR, University of Stellenbosch and the ACRU team.

Potential future tasks involved

- Consolidate past research undertaken on water use by forests
- Identify research gaps and carry out research in these areas
- Set up a reliable monitoring and evaluation (for quantity and quality) system for all dry land crop water use
- Establish a system of knowledge and resource sharing within the forest sector.

LIKELY STAKEHOLDERS RESPONSIBLE FOR CHAMPIONING THE OBJECTIVE

The forestry industry and DWAF will need to jointly champion research into water use by forests. They will need to be supported by tertiary and research institutes and community forestry groups. Input from the water allocation teams in DWAF will be necessary for this task.

There is a range of potential water management options available to the forest sector as described in Section 5.

A comprehensive understanding of these options, their implications, and how they could be applied is needed. The forest sector identified this need because there is a poor understanding of demand management; it is necessary to introduce applied knowledge to the decision-making process; and there is little understanding of the economic tools available or of the consequences on the forest sector of water pricing.

Potential tasks involved

- Carry out research to identify and assess the impacts of the water management options available to the forest sector
- Evaluate the acceptability of options to the forest sector and other water use sectors
- Undertake pilot studies to test potential water management options.

Likely stakeholders responsible for championing the objective

All forest stakeholders will be involved, but the process is likely to be driven mainly by the forest sector. Tertiary and research institutes can offer research support. The NWRS will influence this objective and the appropriate staff in DWAF will need to be involved. Equity, beneficial use and ecosystem functioning are identified in the Act as important goals for water resource management. In addition, the National Forests Act provides for the development of criteria, indicators and standards for sustainable forest management. A research and consultative programme is necessary, because there is little consensus on defining or applying these terms in practice.

Current initiatives:

- The Department of Water Affairs and Forestry has drawn up the policy and procedures for licensing streamflow reduction activities.
- The second edition of Environmental Guidelines for Conservation Management of Commercial Plantations in south Africa, is to be published in July 2000.

Potential future tasks involved:

- Establish an agreed understanding of equity, ecosystem functioning and beneficial use amongst stakeholders
- Set standards for equity, ecosystem functioning and beneficial use
- Develop criteria and indicators to measure the progress and achievement of equity, ecosystem functioning and beneficial use
- Incorporate these criteria into evaluating water use options.

Likely stakeholders responsible f.or championing the objective

This objective will need to be championed by DWAF, as it requires close alignment with national goals. Issues around water quality may require the involvement of the Water Quality Sub-Directorate. Other areas of overlap may be around water allocation, maintenance of the Reserve and the SFRA SEA.

Water demand by a sector or user is not static or absolute at any given time but dependent on external influences. An identification and analysis of socio-economic factors affecting the forest sector, and individual foresters, is an important component of the Strategy. Those external factors which cause an increase in water demand within the forest sector should be targeted. An understanding of these external influences will help ease some of the insecurities felt by the forest sector surrounding change.

Potential tasks involved

- Identify the external influences on the forest sector
- Carry out research to evaluate the impact of external influences on the forest sector
- Establish collaboration with other sectors influencing the forest sector
- Undertake a scenarios and foresighting exercise to understand future trends

- Develop a strategy to allow the forest sector to maintain stability in the light of external influences.

Likely stakeholders responsible for championing the objective

Joint collaboration between DWAF and the forest sector would be useful in meeting this objective. The two groups can bring different perspectives and knowledge to the study.

An appropriate research programme

DWAF, the forestry industry and research institutes already support research into the forest sector. This research must be capitalised on and directed towards WC&DM concerns. There are opportunities for collaboration between DWAF, the forestry industry, and tertiary and research institutions to implement an appropriate research programme on water management within the forest sector. The WRC is currently in the process of developing a research programme to support SFRA and integrated catchment management.

Current initiatives:

- The CSIR, DWAF and WRC has signed an memorandum of agreement to jointly fund a comprehensive research programme to assist with the implementation of the SFRA policy. The research programme has been developed and the first call on research projects has been made.

Potential future tasks involved

- Identify areas where there are gaps in the current information and knowledge bases necessary for WC&DM
- Establish collaboration between the stakeholders in the forest sector to share learning
- Establish a funding mechanism for research
- Commission research in areas where there are information gaps
- Communicate the findings of the research to stakeholders.

The development of a research programme is dependent on all the key stakeholders taking responsibility. This includes DWAF, the forest sector and tertiary and research institutes. It is advisable for the different directorates within DWAF to work together on mutually supportive research. A committee (see 7.4.5) consisting of representatives from the groups may be set up to oversee the implementation of the research programme.

FOREST MANAGEMENT AND USE

VALUE OF THE FORESTS

Timber holds first place among the forest products of New Guinea and Fiji and may do so in the future in the other continental islands, as is generally true on the continents. On the oceanic volcanic islands, however, the native forests bear very few species or individual trees of such size and quality as are suitable for commercial timber and experiments are constantly progressing with introduced plants. The aborigines use wood for house posts, paddles, canoes and sundry minor articles, for which individual trees are selected when needed.

In Hawaii the chief value of the Forest Reserves is in relation to watersheds, because of the demands of the sugar and pineapple plantations and of urban communities and the same probably also holds true for the other high islands under more intensive use.

Where there is mangrove vegetation, it is extensively exploited for fuel and in Fiji is under a 40-year management cycle, as we have seen.

On most of the islands the main use of secondary growth is for firewood and charcoal. For fence posts, highly important in all grazing regions, the naturalized Prosopis chilensis is one of the chief sources of supply. There is no pulp industry of importance, although recent developments in the use of bagasse are significant. Rubber has been tried in the southwestern islands, but not successfully. Minor forest products are of no more than local importance. For the New Guinea native the principal value of the forest lands is as agricultural soil and once the grasses take over, he moves on, or vanishes.

OWNERSHIP AND ADMINISTRATION

Organized forestry, private or governmental, exists in only a few of the island groups. During the Japanese administration of Micronesia, forestry activities were accelerated, with local sawmill operations, minor exports to Japan and a certain amount of reforestation, especially with Casuarina equisetifolia. The present forestry situation in American Samoa and Micronesia has recently been surveyed.

New Caledonia, Fiji (where a Forest Department was established in 1938) and Hawaii have, or have had, some government administration, with Hawaii far in the lead.

In Hawaii a Board of Commissioners of Agriculture and Forestry has existed since 1903. Its activities have centered largely around the establishment of reserves almost entirely for watersheds, the elimination of the hordes of feral mammals, fencing, a very elaborate planting programme that has involved a vast number of alien forest species and the controversial practice of introducing game animals. In the early days, forest plantings were sometimes made in total disregard of basic silvical principles and failures were excessive. For a long time, success was gauged on survival and rapid growth and undoubtedly this is a chief factor in the case of a timber species that is to be harvested.

For self-perpetuating forests, however, this policy may be very deceptive. An intensive report is much to be desired on all these forest plantings of the past, not only with regard to the survival of planted individuals, but also as to whether different species are capable of spreading naturally. Some species may be found to act as "nurse crops" that is, to prepare the way for the introduction of another semi-natural plant community closer to the original self-perpetuating forest.

UTILIZATION

The original uses of the forest were generally unorganized and unplanned. The natives knew their trees and chose carefully the species and individuals best suited for house posts, canoes, paddles, fishnet floats and other articles to meet their daily needs.

The breadfruit was recognized as one of the most valuable of timber trees (propagated vegetatively and existing in endless clonal strains), but was seldom cut, because its fruits were used for food. Coconut also served numerous purposes. Other trees of the strand were selected and many species in the coastalplain forests of the high islands were discriminately chosen, although reports concerning the same species from different island regions vary markedly.

Before World War II organized forestry and forest exploitation can hardly be said to have existed in Oceania. There were a few sawmills on the larger islands, but they served immediate local needs only.

During the war exploitation was stepped up to a high pace to serve the armed forces, especially in the larger Melanesian islands. As in so many tropical regions, the number of species here is great and bewilderingly complex and many of the species have native names only and are unknown to scientific classification.

The wartime operations helped to advance our knowledge of the dendrology and physical characteristics of these forests farther than throughout all previous history, but unfortunately most of such information exists only in unpublished reports and in the minds of the foresters who were directly involved.

RESEARCH

The forestry and agricultural organizations are carrying forward only limited programs of basic forestry research. Usually these organizations are under pressure to serve immediate needs and hence their investigations are oriented toward the planting of exotic tree species of proven worth elsewhere and toward the control of actively destructive factors, such as feral mammals.

On the other hand, scientific research in Oceania is beginning to advance rapidly along general lines and this should benefit forestry in the long run.

Forestry in Oceania involves the management of lands originally or potentially covered with trees and such products of forest land as water and wildlife may locally be more important than commercial timber. For these reasons a sound forestry programme must go beyond the range of conventional silvicultural activities.

It must include interpretations of the vegetation of pre-European and pre-native times and also serious studies of the remaining samples of natural and semi-natural vegetation, which in some instances have very desirable characteristics. The artificially induced forest types, such as those from plantings, should also be investigated with regard both to their continuing or self-perpetuating characteristics and to their values in other respects than for timber production. Because of the extraordinary complexity of the native and introduced floras and because of the wide range of distinctive environmental conditions within relatively small areas, forestry in Oceania offers unsurpassed challenges to the ingenuity of the practical researcher.

FOREST MANAGEMENT ACTIVITIES

To facilitate efficient management, the forest has been divided into three blocks which cover areas of 91, 124 and 129 hectares. The intention is to protect and provide a sustainable and equitable utilization of the forest. Users have been involved in planting of various species assisted by forest technicians and committee members. In 2052 (1992), the women's groups planted 11,000 Sissoo seedlings. The following year the users planted 8536 seedlings of bamboo, Bakaina and Sissoo. There are plans to establish a plantation for Eucalyptus, Sissoo, Khair, Botlle Brush, Gul Mahar, Asoka and other species in the near future. The group is also undertaking NTFP farming in suitable areas under the guidance of forest technicians and other organizations. In collaboration with the Agro Herbal Company, a private company promoting NTFPs which recently set-up a field office in Dipyapuri, around 100 types of NTFP have been identified. The group has also established a tree nursery. The general level of awareness about community forestry has been strengthened and the popularity of the FUG increased. While people were afraid of DFO staff in the past, these relations have now improved.

PROTECTION

The protection wing has the overall responsibility for forest protection and has put in place rules and regulations to achieve this goal. Three user-managed protection posts have been set-up at strategic locations in the forest and three forest guards employed. Their job is to protect the forest against illegal felling, the use of fire and destruction of flora and fauna through grazing, poisoning, shooting and other harmful activities. Controls to regulate the collection of fuelwood, cutting of grass and pruning of fodder are in place and a graded penalty system where fines are calibrated according to the severity of violations is operative. There is a complete restriction on entrance into the forest from Chaitra 1st to end of Jestha as a precautionary measure to prevent forest fires. The group plans to construct a 5 m wide fire line in the north and west of the forest. Notice also that the forest is completely closed for grazing because of the high regeneration rate. The combined effect of these efforts is argued to effectively curtail illicit activities.

FOREST PRODUCT UTILISATION AND DISTRIBUTION

The Forest product distribution wing is responsible for the distribution of forest products and the process governing the allocation of timber is very similar to Dhuseri. Likewise, the forest product rates across the two groups are almost identical. Records are maintained for extracted and distributed forest products. To access timber quotas, users should submit an application along with Rs. 25 and state the reason behind the need for timber. After a recommendation (or rejection) of the application by the head of the sub committee, the committee will allocate timber as per the rules.

The role of the monitoring committee is to evaluate whether users have utilized forest products to fulfill their needs. The monitoring committee has a formal authority to penalise violations. It should be mentioned that while the mechanism for control of timber utilisation in Chautari is a replica of the process in Dhuseri, the above description provides the "official" account of the process of distribution of timber in Chautari. In contrast to Dhuseri, where we know that the control mechanism is ineffective, data limitations make it hard to establish precisely how well the mechanism operates in Chautari. While the incentive problems and scope for arbitrage are as strong as in Dhuseri, it is distinctly possible that the contrast between the official and the hidden economy is less pronounced in this case.

Branches and twigs damaged by the wind are distributed free of cost to users every year from 1st to end of Poush and in Jestha for 15 days but this is restricted to 2 people from each HH. The users can also collect the twigs and fallen small branches every Saturday during Shrawan and Bhadra. Users may also purchase fuelwood from the group for Rs. 100/quintal which exceeds the price of Rs 75 charged by Dhuseri. Grass cutting is allowed from Bhadra 15th to Ashoj 15th and Mangsir 15th to end of Poush for the growth of the tree species. The forest is also open for fodder from Magh 1st to end of Baishak.

Special Provisions and Development Expenditure

Free membership has been granted to schools in the users' area of the CF. Timber is mainly provided to schools against a fixed price. The temple can get up to 5 cubic feet timber free of cost but have to pay for further requirements. The user group has awarded construction grants to Barchuli Junior High School Rajahar for extensions of classrooms and roof support to Saraswati Primary School in ward no. 4. A construction grant of Rs 75.000 has also been given to the sub health post in Rajahar. Other financial and timber support has been provided to other schools, the police office, mothers groups and NGOs. Moreover, a grant for a biogas plant of Rs 75.000, aimed at reducing fuelwood consumption has also been granted. The group has also given support in the form of disaster relief, mainly in connection with Jharahi floods.

Problems, Issues and Conflicts

Dhuseri and Chautari are undoubtedly advanced forestry user groups both in infrastructure development and in their respective approaches to community forestry. The user groups share another common feature: female leadership and participation remains very limited. Moreover, as noted, the motion for differentiation of users by socio-economic status was recently defeated in the General Assembly.

Based on a claim of being deceived by people living near the forest, *e.g.* residents in ward no 8, users from ward 5 have argued that a separate part of the forest should be allotted to them. Having much cultivated land and substantial livestock holdings, ward 5 has a high demand for forest resources to meet agriculture, domestic and livestock needs. The claim of deception was rooted in the observation that users in ward 8, adjacent to the forest, collect more forest products, legally and illegally. In 1999/2000 an interesting conflict emerged over a plan for a ward-wise division of the forest into plots. A demarcation for this purpose was undertaken on the initiative of the then Chairman X. However, users in ward 8, resisted this initiative. Having used the barren areas of the forest for grazing, a ward-wise division of the forest would effectively restrain their grazing opportunities. These users now accused the former Chairman, Mr X of being responsible for conflict claiming that he had received money from people from other wards after encouraging and backing their demands for a ward-wise division.

The committee pays field allowance of Rs 12.-14.000 per year to DFO-staff during the utilization season. The role of the ranger during the harvesting period is to approve various steps in the harvesting process including estimates of felling, blazing and numbering of trees, to grant permissions for sawing in the sawmill as well as permissions to sell any surpluses outside the VDC. This provision of technical assistance falls within their official responsibilities and covered by regular salaries. Despite of this, the FUG is being charged for these services. There is much disagreement about the payment and the scale of payment to the forest officials. The Nepal-German Ayurvedic society was prepared to enter into a 20-year agreement with the user group with a view to promote NTFP production and sales. An almost completed agreement was, however, undermined by strong opposition from the current chairperson Mr Y, and the proposal eventually scrapped.

His official claim was that the project could jeopardise the daily needs requirements of the users, due to a leasing clause in the proposed agreement. Mr X (the then chairperson) who had been supporting the initiative felt that this resistance threatened and undermined his leadership. There is a rather strong element of personal politics in this narrative. Mr Y's hidden agenda was to undermine Mr X's position to overtake the Chairmanship himself, a goal eventually accomplished, since the Ayurvedic society project compelled Mr X to resign. Mr X now claims that he can prove that the present chairman, Mr Y has been involved in illegal activities, more specifically, that Mr Y has felled green trees in conflict with the Operational Plan. However, Mr X is reluctant to provide further details because he is concerned about the reputation of the FUG.

During the chairmanship of Mr Z in 2057/58, an allegation of financial misconduct was raised in the assembly. It was found that he had spent Rs. 11000 to buy alcohol for the DF staff and for employing laborers for the transportation of woods from the forest to the depot. Rs 11.000 was spent on food and travelling to the DFO in order to obtain the harvesting permit. A considerable sum of money was spent to persuade the ranger to approve a larger than previously agreed quota. Another case, accusing Mr Z for misuse of Rs 9000 by giving favour to people already privileged through other FUG-activities, related to the loading of timber. The staff was given Rs 9000 for loading in addition to their regular payments. Poor users argued that they should have been given the loading job and Mr Z was accused of denying the poor employment opportunities.

During his tenure, Mr X provided Rs. 1500/- to a DF staff in the harvesting season, *e.g.* December 2056/57 in connection with a visit to the District Office to obtain the blazing order for felling of trees. Mr X and the then vice chairman requested the ranger to produce the blazing order. The ranger lingered and asked for money. He was given Rs. 1000/- on the spot. The ranger then requested the chairman to buy lunch and incurred further expenses of Rs 500. Two days later the blazing order was sent. After issuing the blazing order, DF staff marked 1700 cft for felling. While cutting it was discovered that some trees were hollow and the committee asked for permission to cut more trees to fill the allotted quota. The ranger accepted this request after extensive bargaining. Having been granted this second permission, the majority of the committee, who were aligned with the Chairman and with forest staff overstepped their mandate and felled old green trees as well. Inspecting the spot, the ranger gave the impression that the illegal felling had been reported to the district officer. He said that the committee should attend DFO for clarification without any further delays. The committee members were worried and asked the ranger to do whatever he could to minimize the offence. For this favour the ranger demanded 20 cubic feet of timber. The committee provided the timber immediately to save themselves from further trouble and embarassment.

Another interesting example concerns the sawing of timber. In July 2002, the ranger had given the committee permission to saw in Dibyapura Saw Mill at a rate of Rs. 38/cft.[3] Instead, the committee decided to process the timber at Pragatinagar saw mill (4 km west) which offered a rate of Rs. 30/cft, thereby permitting a saving of Rs. 8/cft. When informed the ranger sent a letter asking for clarification. He didn't approve of the answer thinking he had lost commission from the saw mill. The committee members involved in this incident are reluctant to provide further information about the matter.

MANAGING FORESTS FOR CLEANER WATER FOR URBAN POPULATIONS

Access to clean water is one of the most fundamental of human rights, but currently more than one billion city-dwelling people lack access to clean water. Generally it is not because water supplies are insufficient. Rather, this crisis is due to an inability to organize supply properly to meet demand. This failure is particularly frustrating in that nature contains the necessary mechanisms to provide clean, healthy water, including the filtering effect provided by healthy forests in watersheds. Yet in many parts of the world environmental mismanagement has led to a critical shortage of freshwater.

This article highlights how some of the largest cities in the world are able to supply sufficient freshwater to their inhabitants at least in part through the protection of forests. It identifies some key policy initiatives that could help reduce the vast number of people whose lives are dominated by the daily search for clean and safe water.

SUPPLY AND DEMAND

Water is, in theory, a quintessentially renewable resource. Water covers most of the world's surface, and over much of the world it falls unbidden from the skies. Yet because of the carelessness and profligacy with which water resources have been used, the speed of human population growth and the increasing demands for water, the provision of adequate, safe supplies of water is now a major source of concern, expense and even international tension.

The poorest members of society, unable to afford safe water, suffer the greatest impacts. One in five people in the developing world lives without a reliable water supply. Lack of clean water has dire short- and long-term health impacts including increased infant mortality and impaired ability to work, which reduce industrial productivity and put pressure on already overstretched health services. On the other hand, access to clean water can have dramatic positive health impacts. For instance, it is estimated that when clean water is available the risk of early death is reduced by 23 percent in Uganda and 30 percent in Cameroon (UNDP, 2006).

Today, around half the world's population lives in towns and cities, and of these people one-third live without clean water or adequate sanitation (United Nations Human Settlement Programme, 2003). Municipal authorities have a variety of ways of supplying drinking-water, depending on where they are located, the resources available, social and political issues and the willingness of the population to conserve water. Most cities rely on the collection and diversion of existing surface and underground freshwater sources. Only minor amounts, on the global scale, are extracted directly from rainwater or from the seas. Until recently, the main focus of efforts to improve urban water supply has been within the cities themselves, including better distribution systems, treatment plants and sewage disposal. However, many authorities are now increasingly looking at land management systems that can help maintain pure water at the source.

WATER QUANTITY, QUALITY AND REGULARITY

People have settled historically in areas rich with natural resources, and today most of the world's population lives downstream of forested watersheds (Reid, 2001). Societies have created strong cultural links with forests, and it is widely assumed that forests help to maintain a constant supply of good-quality water. Conversely, loss of forests has been blamed for problems ranging from flooding to aridity.

In fact, the hydrological role of forests remains a subject of debate. Some of the common assumptions about the benefits that forests bring are wrong in most situations; for example, most forests do not increase water flow in a catchment (in fact the reverse is often the case), nor do they necessarily control flooding. On the other hand, some positive benefits, including particularly their potential to supply relatively pure water, are frequently overlooked. Impacts of forests are influenced by many factors including the age and species of the trees, the amount of watershed under forest, soil, climate and forest management practices. A meta-study conducted for the World Wide Fund for Nature (WWF) on the role of forest protection in drinking-water provision (Dudley and Stolton, 2003), including a survey of more than 100 of the world's most populous cities, revealed – as described below – a clear link between forests and the quality of water coming out of a catchment, a much more sporadic link between forests and the quantity of water available and a variable link between forests and the constancy of flow.

Quality

Forested watersheds generally offer higher-quality water than watersheds under alternative land uses, if only because virtually all the alternatives – agriculture, industry and settlement – are likely to increase the amounts of pollutants entering headwaters. Quality can also be higher because forests sometimes help to regulate soil erosion and reduce sediment load, although the extent and significance of this function will vary. Undisturbed forest with understorey, leaf litter and organically enriched soil is the best watershed land cover for minimizing erosion by water. While forests are less able to control some contaminants (the human parasite *Giardia lamblia,* for example), in most cases the presence of forests can substantially reduce the need for treatment for drinking-water and thus radically reduce costs of supplying water.

Where municipalities have protected forests for their water resources, quality issues have generally been the primary motivation. In Tokyo, Japan, for example, the Metropolitan Government Bureau of Waterworks manages the forest in the upper reaches of the Tama River to increase the capacity to recharge water resources, to prevent reservoir sedimentation, to increase the forest's water purification capacity and to conserve the natural environment. In Sydney, Australia, the Catchment Authority manages about one-quarter of the catchment as a buffer zone to stop nutrients and other substances that could affect the quality of water from entering storage areas.

Quantity

The situation with regard to the flow of water from catchments is more complex. Despite years of catchment experiments, the precise interactions between different tree species and ages, different soil types and management regimes are still often poorly understood, making accurate predictions difficult. In contrast with popular assumptions, many studies suggest that in both very humid and very dry forests evaporation is likely to be greater from forests than from land covered with other types of vegetation; thus less water flows from forested catchments than, for example, from grassland or crops (Calder, 2000). The evidence seems to suggest, however, that cloud forests (Bruijnzeel, 1990) and some older natural forests (such as old*Eucalyptus* forests) can increase net water flow. Some cities factor management of these forests into plans for maintaining adequate water supplies.

Regularity

Constancy of flow is as important as total quantity, in terms of both maintenance of dry-season flow and absence of flooding during periods of heavy rain. Here opinion remains divided, as examples of very different responses can be found. In some cases dry-season flow is depressed by the presence of trees, while in other cases it is increased. Natural forests and plantations have different effects, but again these do not show a constant trend. In very general terms, forests often help to regulate relatively minor floods but are seldom able to prevent occasional, very major floods. Flooded forests – both lowland forests such as the Várzea forests of the Amazon and swamps in the uplands – have a more definite role in regulating water supply. However, the debate about the role of forests in maintaining constancy of water flow continues; a recently published study suggests that natural forests have a larger role in flood prevention than has generally been argued of late (Bradshaw *et al.*, 2007).

URBAN SUPPLY

The contributions of forests in providing clean water depend to a large extent on individual conditions, tree species and age, soil types, climate, management regimes and needs from the catchment. It is therefore perhaps not surprising that information on best practices for policy-makers remains scarce and models for predicting responses in individual catchments are at best approximate. Towns and cities are faced with a bewildering diversity of opinions on which to make hard financial and politically charged decisions about their water supply. Yet many of the world's biggest cities are choosing to rely at least in part on forested areas to help maintain water supplies.

The meta-study described above (Dudley and Stolton, 2003) indicated that about one-third (33 of 105) of the world's largest cities obtained a significant proportion of their drinking-water directly from forested protected areas. At least five other cities in the review obtained water from sources originating in distant protected forested watersheds, and eight more obtained water from forests managed in a way that gave priority to their functions in providing water. In a number of cases there is also good evidence that forests help maintain water flow – for example in Melbourne, Australia and in some cities fed by cloud forests such as the Caribbean National Forest in Puerto Rico. However in some other cases where cities have been protecting forests specifically to maintain water supplies, there is little hard evidence that forest protection has this effect.

Many municipalities (although certainly not all) cite maintenance of a pure water supply as a reason for introducing forest protection or reforestation. In the United States, all states are required under federal law to have a Source Water Assessment, which promotes the idea that protecting drinking-water at the source is the most effective way of preventing drinking-water contamination (NRDC, 2003).

The city of New York is famous for its use of protected forests to maintain its high-quality water supply. This approach was supported by popular vote in part because it was a cheaper option than building more treatment plants. Other cities in the United States also rely on forested catchments.

Around 85 percent of San Francisco's drinking-water comes from the Hetch Hetchy watershed in Yosemite National Park. In Seattle, Washington, the primary sources of water are the Cedar River watershed and the South Fork Tolt watershed, which together serve a population of 1.2 million people with unfiltered drinking-water.

MANAGING FORESTS FOR WATER

Forests offer a range of options for water provision, depending on their type, location and age and on what users need. Cities may choose a number of different forest management options, including protection, sustainable management and, where necessary, restoration. Those responsible for water supply and forest management are faced with a number of questions: about whether forested watersheds offer real benefits for water supply; if they do, how much forest is required to gain these benefits; and how forests in watersheds can best be managed to protect water supplies. In most cases decisions will need to be taken in a context of competing demands on land, so that management for water will have to be balanced and traded off with other uses.

The following questions all need to be answered before any decisions are made about forests managed for water.

- *What are the most pressing needs regarding water supply?* Are the pressures on water supply primarily driven by the need to get enough water, or a constant supply of water, or is the priority more to do with water quality? What quality issues are most important? For example, sediment will be most important for hydropower, whereas pollutants such as agrochemicals will also be of concern for drinking-water.
- *How is vegetation in the catchment likely to affect water quality and flow?* This needs specialist analysis, although some general points can be made. For example, cloud forests are likely to increase water, some old natural forests may also increase flow, and young forests and plantations are likely to decrease flow. Individual cases need to be assessed, depending on soil, climate, forest types and age, and management regime.
- *What is the land use?* Current status is important, but so are recent changes and likely future trends. Answering these three questions will help to determine what natural vegetation (and perhaps other land uses) in the catchment offers in terms of water supply and whether future changes are likely to increase benefits or create problems. With this information, more strategic analysis can help to plan optimum management interventions:
- *What other demands are there on land in the catchment and how much might be available for water management?*Are other pressures on land likely to improve or degrade water? How much land is available, partially or completely, for water management? Can current land uses be improved from the perspective of the water from the catchment? What impacts would these changes have for local people and what are their needs and wishes? Can catchment areas also be used for other land uses, such as recreation or biodiversity conservation?
- *What are realistic management options?* Present and future management options should be analysed, including establishment and maintenance of protected areas, forest restoration and other forms of land use.

The analysis should tell whether the presence of forests can help supply the water required from the catchment and provide the information needed to make informed choices about a landscape mosaic that will fulfil both water needs and other needs from the watershed.

VALUING FORESTS

In many cases the economic case for managing ecosystem services can provide the impetus for sustainable forest management. A team of researchers from the United States, Argentina and the Netherlands put an average price tag of US$33 trillion per year on global fundamental ecosystem services, which are usually taken for granted because they are free. Water regulation and supply was estimated to be worth US$2.3 trillion (Costanza *et al.*, 1997). At the national level, the economic value of the water storage function of China's forests is estimated as 7.5 trillion yuan (approximately US$1 trillion), three times the value of the wood in those forests. Another study calculated that the presence of forest on Mount Kenya saved Kenya's economy more than US$20 million by protecting the catchment for two of the country's main river systems, the Tana and the Ewaso Ngiro (Emerton, 2001).

The issue for policy-makers is how to translate these values to help support particular types of land management. One reason why it has proved so difficult to halt and reverse global forest loss is that those who manage forests typically receive little or no compensation for the services that forests generate for others and hence have little incentive to manage them sustainably. Even when areas are protected, values such as water provision are often not recognized by the users. Owing to the serious financial difficulties faced by protected areas in Venezuela, in 1999 the Instituto Nacional de Parques(INPARQUES), the State agency for protected areas, considered charging water companies for the direct services they obtain from these areas (including the three protected areas that are the source of water for the country's capital, Caracas). However, until now this initiative has not been further developed (Courau, 2003).

Recognition of this issue has encouraged the development of systems in which land users are paid for the environmental services that they generate through management. The central principle of the "payment for environmental services" (PES) approach is that those who provide environmental services should be compensated for doing so from those who receive the services. Projects using water resources as a springboard for PES schemes have mainly been developed in Latin America, but interest is quickening throughout the world. In Quito, Ecuador, for example, water companies are helping to pay for the management of protected areas that are the source for much of the capital's drinking-water.

7

Challenges For Sustainable Arid Zone Agroforestry

Agroforestry has been followed since time immemorial for sustainable livelihood in the arid zone, but in the recent past some changes have occurred, due to which the need is being felt to develop agroforestry systems, which may be able to meet the requirements of sustainable livelihood in the future.

HEAVY BIOTIC PRESSURE

Chalcolithic and Iron Age maps of the Indian subcontinent suggest that human occupation of the current Thar Desert region was thin during these earlier ages, and that populations depended upon hunting and limited pastoralism. From the analysis of available census records, the population of the Thar Desert region registered an increase of 490 per cent from 1901 to 1991. The present population of the Thar is about 40 million.

Decennial variation in population from 1901 to 1991 in the Thar showed a growth rate of 186 per cent (between 1901 and 1971) as compared with 132 per cent for the whole country. The population growth rate for the decades 1971-1981 and 1981-1991 were 36.7 per cent and 30.7 per cent, respectively. The population density is quite high compared with the global average of 6-8 persons/km^2 for arid zones. The major factor responsible for such phenomenal growth of population in the Thar is a wide gap between birth and death rates. Moreover, social values favour large families and positive sanctions for fertility outnumber negative ones.

Cropping on Marginal Lands

Of the total area of the Thar, only 1.3 per cent is now under forest cover. The net cultivated area increased by 54.1 per cent from 1915 to the early 1980s. This trend in cultivated area increase has further accelerated during the past decade (the 1990s), having increased by 11.7 per cent compared with the amount in the early 1980s. As a consequence of this, barren, cultivable and uncultivable wastelands, permanent pastures, and fallow land have declined substantially.

There was a significant decline in grazing land area as marginal lands were brought under cultivation in recent years (the late 1990s). Because of low and erratic precipitation, restricted to a very short period, only one crop, the kharif (monsoon) season crop, is sown each year in most of the Thar.

The common crops are pearl millet, cluster bean, green gram, moth bean and sesame. In the arid zone of peninsular India, in addition to these crops, sorghum, maize pigeon pea and other coarse millets are also grown.

It has been observed that the agricultural holdings during late medieval times were only one fifth of those found today. By 1951, much of the potential agricultural land was brought under plow, and the trend since then in already established agricultural tracts has been intensification of farming, and the expansion of agriculture into new lands that may be considered environmentally marginal.

Mechanized field operations (use of tractors) not only increase soil erosion but also reduces the naturally regenerated population of various useful species.

Unsustainable Development Policies

With the advancement of science and technology, efforts have also been made to enhance production in the arid zone. Some of the technologies like sand dune stabilization, wind-breaks and agri-horticulture have produced good results because they are developed in synergism with nature (ecosystem) and natives (people and vegetation).

But some other technologies imported from less environmentally stressed areas have proven to be unsustainable. For example use of tractors adversely affected the natural regeneration of native trees like Prosopis cineraria, and increases wind erosion. Irrigation with waters from the Indira Gandhi canal network has caused problems of water logging and, secondary salinization. Several exotic species either become invasive, like Prosopis juliflora, or didn't prove economically productive, like Faidherbia albida, jojaba etc. The overall impact of these imported technologies, along with high biotic pressure and a fragmented approach to desertification control, can now be observed in terms of loss of biodiversity, increasing desertification, and other adverse impacts. It is now imperative to think broadly and plan holistically for the future of arid zone agroforestry, the basis of sustainable livelihood in the arid zone. Throughout the arid zones, there is no dearth of problems, but rapidly increasing desertification (some call it land degradation) is a worldwide problem. Land degradation appears to be more of a descriptive term than desertification, as it better describes the process of wind and water erosion, soil salinization, and loss of vegetation cover. It helps in clarifying relationships among climate, topography, soil structure, land use, hydrology, and vegetation. Throughout the arid zones of the world, the major factor responsible for land degradation is soil degradation.

At the world scale, the GLASSOD database indicates that in arid zones, more than 60 per cent of soil degradation is caused by wind erosion, with human-induced water erosion accounting for 8 per cent of the phenomenon. At the global level, 349.6 million hectares of land in arid zones are characterized by a light to moderate degree of soil degradation, and 42.9 million hectares have strong to extreme soil degradation.

Besides wind and water erosion, over-grazing, high magnitude agricultural degradation activities, over exploitation of vegetation for domestic use, *e.g.* for fuel, and faulty land use are other factors responsible for ever increasing land degradation in arid zones.

The links between land or soil degradation and vegetation in arid zones are complex. Vegetation and soils display differing resilience to disturbance. Although in arid zones vegetation communities can be readily disturbed and degraded, recovery rates are very slow in comparison with those of other ecoclimatic conditions. Vegetation in arid zones is highty sensitive to natural climatic variability. Decreased production and an ability to lie dormant are two strategies displayed by arid zone plants for survival at times of moisture deficiency. Although an adequate database is not available, two useful relationships between soil degradation and vegetation can be determined. First, vegetation cover may help to indicate the vulnerability of a particular habitat to soil degradation, and second, vegetation cover or community type or plant species may indicate that soil degradation.

SUSTAINING AGRO-BIODIVERSITY

Nations were required to evolve a *sui generis* system for varietal protection, go in for patenting or adopt a combination of both, but not in isolation from the broader goals set out in agenda 21 of the CBD. Such a *sui generis* system, in absolute sense, provides an institutional mechanism and legislative protection to manage and enrich *inter alia* the agro-biodiversity, within the national territory. It would give due opportunities to on-farm conservation for sustainable use besides protecting and safeguarding the interests of those who conserve and maintain the vital components. Issues that the nations must consider, as per the requirements of CBD, include to plan and implement in an effective legal and/or regulatory frame work which would ensure no harmful effects even on territory beyond national borders upon consequent use of various components of agro-biodiversity; the regulation of exchange of germplasm is done in conformity with the charter of the UN and the international law, and that an intellectual property protection is provided to plants or plant parts in other member countries, etc.

Nations may be required to maintain two contrasting agro-ecosystems. First, the traditional agro-ecosystems encompassing and protecting biological diversity therein along with its biotic, abiotic and environmental components providing/maintaining due opportunities to sustainability and the micro-evolutionary processes. The cost of maintaining such traditional agro-ecosystems, through imparting awareness and providing suitable compensations to the local folk who would practise it, is bound to be high but looking into the indirect impacts of such conservative mechanisms particularly on the climatic equilibrium, the CBD appealed to Global Environmental Fund (GEF), in one of the decisions made in the third meeting of the Conference of Parties (COP-III), to make liberal financial grants for the above purposes pending settlement of the issue of establishing a Global Biodiversity Fund, which is being strongly advocated by several partners and fora. Lessons could also be learnt from the more recent indigenous knowledge and social framework of traditional monospecific cropping systems so as to help maintain the ecological equilibrium in advance, highly productive, intensive agro-ecosystems. Funding provisions may specifically be made for conducting research on these aspects.

Sustaining Alternate Ecosystems

A concern has been shown that the traditional agro-ecosystems should be conserved along with diverse evolutionary forms of cultivated plant species as well as farmers' conscious selections, particularly in the areas of origin and diversity of crop plant species and their relatives. This, obviously calls for concerted efforts towards developing a better understanding in terms of the following:

- Classification, indexing and inventorisation of traditional agricultural systems, still in vogue, with respect to their ecosystems, crop combinations, seasons and/or systems of cultivation;
- Potential core area(s) under each, respective system or sub-system; and
- The custodian community.

At the same time, guiding and maintaining diversity in the advance agro-ecosystems through research, extension and legislation is considered essential to tackle several problem situations, such as, biotic virulences, abiotic stresses and adverse climates. Rapid appraisal systems are, therefore, essential so as to maintain and guide requisite diversity in these agricultural systems to enable fast detection and appropriate replacement of the varietal components, such as those succumbed to diseases/pests/abiotic stresses or those which have been surpassed by more productive types in research. The creation and maintenance of agro-biodiversity could be ensured through multi-crop, multi-varietal mosaic cropping which shall mimic the co-existence of cultivated plant species and the within-species diversity in the traditional agro-ecosystems. Also, the maintenance of the latter through awareness generation and providing suitable compensation to the local people who practise it, would be essential for an effective management of agro-biodiversity.

In Situ on-farm Conservation

There is a demanding need to enumerate agreed scientific principles and practices for 'on-farm' genetic resource conservation. It includes usable classification and nomenclature systems for landraces, as well as methodology for their characterisation. Due to lack of available information of this type, it is imperative that studies should be carried out on these aspects side-by-side while implementing 'on-farm' conservation programmes. A better understanding of the temporal and spatial movement of genes and varieties into and out of cropping systems and the causes of loss or recession of some varieties or traits over others should be highly useful in further setting the pace and direction of crop evolution. Adequate studies of some of the traditional practices and customs have not been carried out in most of the agro-biodiversity rich areas. Systematic surveys and collection of data on the traditional knowledge base, including methods to maintain healthy populations and seed, require an inquisitive follow-up and indexing.

The 'on-farm' conservation is generally associated with traditional agro-ecosystems and small scale farming although the farmers' participatory approach has been more advocated to help select suitable high yielding varieties (HYVs) for the advance cultivation systems. The large-scale commercial farms also have options to promote agro-biological diversity through crop rotations, intercropping, cover crops, green manures, paddy-cum-fish culture, apiculture, etc.

The seed requirements for such situations would depend upon the adaptability, preference and acceptability of component HYVs for each sub-situation. Such practices need to be standardised as per diverse agro-ecological zones, so that sustainability of seed component is ensured under a dynamic intensive cultivation system. The more likely negative impact on sustainability of seed under the intensive cropping system may occur due to successively superior varieties being evolved at frequent intervals, the imminent displacement of modified agro-ecosystems which were created by cultivating such superior HYVs, the evolution of virulences or other environmental factors, such as, off-site pollution or accumulation of certain harmful gases like methane or ozone, etc. which need to be addressed by well planned, self-generating mechanisms.

Domestication of individual crop plant species is unique to agro-ecological sub-regions and localities and corresponds to cultural and ethnic diversity, including food habits, habitats, climates and available crop plant species. There is a purposeful need to maintain their representative seed 'on farm' because re-building new varieties for similar situations with same amount of adaptability and phenotypic plasticity would require strenuous planning, breeding efforts and resources, including funds and time. Capacity building for 'on-farm' conservation of traditional 'types' requires formulation of suitable policies, sectoral and cross-sectoral plans and programmes for implementation and also interim financial mechanisms. Recognition of the associated local knowledge, innovations and informal practices would be needed to promote traditional farming systems, whereas establishing community seed banks in cooperative or private sectors, developing and maintaining cultivar/landrace registration system, etc., is required for smooth replenishment. The recognition of 'Farmers' Rights' is important along with formulation of implementable laws and regulations so as to ensure availability of the much needed 'seed' with due compensation or assurance of benefit sharing to its holders. Country level implementation of Registration of Plant Germplasm has been implemented in India by the ICAR and the planning/action related to other aspects is also initiated.

Compiling case studies for conservative, low external output based production systems through incentives for developing environmentally sound technologies for conservation, replenishment and sustainable use of biological diversity are issues which lack direction and, therefore, need sincere input from all concerned under the *sui generis* situations.

Geo-sectoral Approach

Contiguity of adjoining areas across the geo-political barriers, representing similar genetic diversity for particular species genepool, can be attributed as a basis to conserve, utilise and share the benefits of the *in situ*-on-farm conservation. Safeguarding the regional 'gene' interests could be ensured through sub-regional consortia and regional networking. The increasing role of Global Forum and the regional fora in strengthening relationships among the national agricultural research systems (NARS) and between various NARS and the Commonwealth Group of International Agricultural Research Institutions (CGIAR), could be viewed as new initiatives in desired direction which need to be pursued. Technological and technical needs can be gradually fulfilled for innovating, standardising and researching into the rapid appraisal systems for agro-biodiversity management using and strengthening the information network(s), as well as database system(s).

Due to the recent global interest for germplasm exchange and concern for protection of Plant Breeders'/Farmers' Rights there is urgency that the countries develop a regional consensus by way of the following:

- By entering into agreement for undertaking joint explorations in the region for collecting genepools of crops for which the region is centre of origin;
- By adhering to effective plant quarantine so as to maintain an effective mechanism for exchange of risk-free germplasm;
- By developing *sui generis* methods of protecting the breeders' and farmers' rights by each country considering its own unique situation

such that the benefits accruing thereof are also available to the other nations of the region; and

- By instituting funding mechanism and ensuring financial support for the regional PGR activities. It is pertinent to emphasise that the seed industry be facilitated and the security of certain vital components ensured.

Consequent upon the adoption of CBD and also implementation/enforcement of General Agreement on Tariffs and Trade (GATT)/World Trade Organisation (WTO) provisions, new issues and challenges of global significance have opened up. Accordingly, the sovereign rights of the nations on their bio-resources have been specifically recognised. As a consequence, it has become responsibility of the nations to conserve these resources and ensure their availability for their systematic utilisation on sustainable basis. Coupled with this, an equitable sharing of benefits accruing from the use of plant genetic resources (PGR) is an important issue that influences the other vital issue of access to PGR. The funding support from GEF has yet not set its pace which emphasises the need to set up a Global Fund which could *inter alia* help shape and refine the action points on seed security and the seed industry. There is an urgency to address the identified areas in harmony with the other institutional mechanisms in the fray. Recognising that the gene rich countries have a potential to release naturally evolved variability suited to several diverse situations, there is an adequate need to match with a generous global commitment towards the funding requirement to sustain the traditional/naturalised diversity on-farm at certain representative sites located within the diversity-rich areas of the world. Some activities might include, but not limited to:

- Regional exploration teams comprising experts from countries involved could be assigned survey, collection and bio-prospecting of agro-biodiversity. Whereas the hitherto unexplored bio-resources should be collected, evaluated, documented and shared for immediate use besides long-term conservation in the gene bank, the well adapted and popular cultures should be identified along with their companion crops/varieties and the cropping systems, for their on-farm conservation;
- Seed multiplication, testing and distribution programmes should be launched for the traditional varieties/folk cultures/landraces which have edge over other co-cultivated variable forms and the farmers provided with critical inputs, either as subsidised or free-of-cost;
- Farmers should be encouraged to plant 'rest of the traditional cultures' in a small block of their cultivation field by ensuring seed of all the diverse component cultures from the agency/community seed bank. The produce should be re-purchased at the premium rate of seed rather than commodity;

- In order to develop regional capabilities/capacities, the human resource orientation and development needs to be given top priority, giving equal weight to educating both genders;
- Existing gene bank facilities should be utilised for long term conservation of seed representative of diverse types for posterity. The satellite, medium term facilities, on the other side, should be used to maintain active forms with a view to replenish the lost or diminished types in a catastrophe/crisis; and
- Exchange of germplasm seed needs to be regulated by strict phytosanitary regulations and quarantine measures so as to minimise risk from new pathogens/pests prevailing outside the region. Joint ventures in bioprospecting, biosafety and biotechnology, including cryopreservation, *in vitro* conservation or handling of transgenics/ genetically modified organisms (GMOs) need to dwell upon principles of sustainability and synergies/mutual cooperation. All such activities require adequate funding support which should be catalysed and facilitated as per need of the hour.

The damage being done to land, water, flora, fauna and the atmosphere altogether poses the challenge of future food security. It is the human interference that created the crop plants from the wild to their present day commercial forms. Their agro-biological features and dispersal mechanisms have changed drastically since then. The commercial forms are absolutely dependent on human beings for their perpetuation and survival. It is, therefore, appropriate to look into the interests of the seed producers of the required/ desired types. The management approach would require monitoring of the varietal component at frequent intervals so that the balance could be maintained between the interests of the seed producers and the phyto-ecological requirements of the varietal components. Involvement of both governmental and non-governmental bodies and the individuals, as also the genetic resource and patent literacy, constitute an essential elements of systematic sustainability and developmental plans.

ARID ZONE AGROFORESTRY

Drylands cover more than 60% of the earth's surface. Arid zones are described as a part of the drylands, and have the most severe edapho-climatic conditions. The major distinguishing feature for defining and for planning for the arid zone is the low rainfall (below 500mm or Aridity Index <0.20) with more than 50% inter-annual variability. This makes a great difference in terms of the nature of the ecosystem, the socio-economic environment and the challenges for sustainability. High wind and solar regimes further increase the effect of rainfall variability and the whole complex makes a fragile ecosystem in which small disturbances may cause great loss to the sustainability, which are sometimes irreversible. The hot arid zones of the world are economically and environmentally disadvantaged, with unique problems.

The ecosystems of these zones are highly fragile with large risks that cause severe impediments to development programmes. Of the total land area in the world, arid zones cover 18.8%. The hyper-arid zones account for 22.3% of the total arid zones. These arid zones are diverse in terms of climate, soils, vegetation, animals, and the lifestyles and activities of the people. Little but variability in rainfall and the presence of distinctive periods of drought are the characteristics of the arid tropics. Often, the terms drought and aridity are used incorrectly. A drought is a departure from average or normal conditions in which shortage of water adversely impacts on the functioning of ecosystems, and the resident population of people, whereas aridity refers to the average conditions of limited rainfall and water supplies, not to the departures there from. In general, arid zones are characterized by pastoralism and little farming but there are always exceptions.

Arid zones are found in almost all the continents of the world. In North America, arid zones are found in the southern US and Mexico, and in South America they are located in Brazil, Argentina, Chile, and in some other countries. In Africa, substantial parts of Saharan Africa, Ethiopia, and Namibia, are classed as arid environments. The hot arid regions of India lie between 24° and 29° N latitude, and 70° and 76° E longitude, covering an area of 31.70 million hectares, and involving seven states: Rajasthan, Gujarat, Punjab, Haryana, Andhra Pradesh, Karnataka, and Maharashtra. In total, 11.8% of the country is under a hot arid environment. The arid regions of Rajasthan, Gujarat, Punjab, and Haryana together constitute the Great Indian Desert, better known as the Thar Desert, which accounts for 89.6% of the total hot arid regions of India, and constitutes the principal hot arid zone of the country.

Table : Distribution of arid regions in different states of India.

State(s)	Area	Percent of
(10^6 hectares)	total	
Rajasthan	19.61	61.0
Gujarat	06.22	19.6
Punjab and Haryana	02.73	09.0
Andhra Pradesh	02.15	07.0
Karnataka	00.86	03.0
Maharastra	00.13	0.4
Total	31.70	100

In arid zones, vegetation is typically sparse, and is comprised of perennial and annual grasses, other herbaceous plants, shrubs and small trees. The native plant species have adaptations that enable them to reproduce, grow and survive in the most inhospitable edapho-climatic conditions. Some plants have evolved special root systems, while others have unique leaf characteristics that allow them to withstand prolonged periods of drought. The number of tree species is very limited in arid zones, and in general, they are very slow growing due to limitations of environmental conditions, but nowhere in the world are they are so intricately associated with the life of human beings.

To evade or minimize the adverse affects of frequent droughts, the native peoples in arid zones have often developed production systems in which woody perennials have a very important role, both from a productivity as well as a resource conservation point of view. Their centuries-old experiences, under diverse socioeconomic and cultural conditions, were passed from generation to generation and people have established well-contained systems of production (silivipastoral/agrisilvicultural/agrisilvipastoral/livestock husbandry) that accrue maximum benefits from their woody components. When arid zones have been looked at from the perspective of forestry development, the focus has been on the management of trees and shrubs that are either native to a particular arid zone, or have been introduced, especially for conservation purposes.

However, concern for trees and forests has increased in arid zones as it has in many other regions in recent years. This concern has been fuelled mostly by the disappearance of large areas of woodlands and trees from such fragile ecosystems in many countries of world. In the arid zone, closed forests are seldom available. In India, traditional animal husbandry and agroforestry practices have been used to manage parklands, rangelands, reserved silvipastures near holy places, lay farming, and run-off farming (traditional watershed management). Trees are managed mainly for their non-wood forest products (NWFP) and their environmental services. Animals are an essential part of the production system. The whole system has been traditionally developed to spread the risk of drought in diversified components and efficient utilization of scarcely available natural resources.

ARID ZONE AGROFORESTRY AND FOREST MANAGEMENT

As stated above, the prevalent form of forest management in arid lands is agroforestry, which is concerned with the management of trees and shrubs for conservation and sustainable development. Agroforestry in arid lands differs in many ways from the timber forestry practiced in moist tropical and other temperate ecosystems and in many cases its applications are broader in their scope. These applications include production of wood for fuel and poles; horticultural practices for fruit and nut production; range management for fodder and forage production; forestry to modify microclimate for improved agricultural crop production; and protection forestry on land susceptible to wind and water erosion.

Many rural people combine these applications of agroforestry with other land uses linked to their needs and social values. The value of forestry in the arid zone is difficult to evaluate in monetary terms. Arid zone agroforestry therefore needs to be understood broadly as management of trees and shrubs to improve the livelihood and quality of life for rural people in the arid environment. It is in this diverse context that policy developemnt must take place if sustainable development of forest resources in the arid zone is to be achieved.

Agroforestry is being followed traditionally in the arid zone to minimize the risk of drought and sustain life in climatic adversities. Several need-based systems have developed traditionally and some have been improved or incorporated in recent times.

FUTURE DIRECTIONS FOR ARID ZONE AGROFORESTRY

Every ecosystem has some buffering capacity that enables it to remain unchanged for some time if it is overexploited.

While this applies to the arid zone, its fragile nature limits its buffering capacity. Thus, for a few decades the shortcomings of policy were not observed but as the buffering capacity is gradually being eroded, the indicators of unsustainability as a whole, *e.g.* decreasing natural resources, deforestation and desertification, become visible. For fragile ecosystems, like those that occur in arid zones, one thing needs to be recognized by policy makers and scientists, that is; sustainability of agroforestry in the arid zone depends on the degree of synergism with '3Ns' *i.e.* Nature, Native vegetation and Native people.

Some of the future directions for arid zone agroforestry need to focus on the evaluation, improvement and encouragement of sustainable traditional agroforestry systems, exploring the possibilities of natural regeneration/rehabilitation of arid zone ecosystems, NWFPs as the basis of sustainable livelihood, rethinking of research priorities, and policy improvement. Evaluation, improvement and encouragement of sustainable traditional agroforestry systems that have been developed for centuries in harmony with nature. Agroforestry systems have been developed and used in this region since time immemorial. This may be because the inhabitants of this region found that no other option was sustainable aside from conserving and harvesting whatever nature is giving. The present forms of traditional agroforestry developed over the centuries through the experiments of earlier generations in all types of climatic variations. The validation of their sustainability is very difficult to establish by conducting short-term experiments lasting only 5-6 years. Thus, there is immense possibility for evaluation, improvement and encouragement of these traditional agroforestry systems.

Exploiting the possibilities of natural regeneration/rehabilitation of arid zone/arid zone agroforestry as an alternative to failed plantation/rehabilitation strategies. It is a mechanism of restoration or maintenance of ecosystems by nature that trees produce numerous seeds. A small part of this seed population is sufficient to replace the loss of plant population due to various factors. Human beings propagate the plants artificially to compensate for what they harvested form the natural vegetation or to get a higher quantity of the natural products. Those needed in large quantities have already been domesticated, *e.g.* food crops, fruit trees etc. Still a large number of natural species are wild or depend on natural regeneration. Trees in agroforestry may fall under this category. Natural regeneration as well as artificial tree establishing technologies works comparatively better in humid and irrigated areas. In the arid zone, due to climatological, physiological and sociological reasons artificial establishment has not achieved the required success.

The need for domestication and technology development for sustainable harvesting and value addition for non-wood forest products (NWFPs). Unlike the humid or sub-humid regions where trees are typically grown for timber, the trees are grown in this arid region mainly for non-wood forest products (NWFPs). In the arid zone closed forests are seldom available and agroforestry has been followed since time immemorial in which people conserve and manage whatever seeds of native tree species like Prosopis cineraria and Zizyphus numularia germinate in the crop field or on the common land. After the establishment of trees only annual lopping of branches is done which provides precious fodder during scarcity/drought. Taking the example of P. cineraria, this tree is found in this region at densities ranging from 40 to 120 trees/ha.

This we can say is the natural forest in common or private lands otherwise no dense forest is found as in the high rainfall areas. People lop the branches of this tree in the month of December and it becomes lush green in the month of May-June when no other fodder is available. Simultaneously, pods develop which are used as vegetables. Similarly other tree species provide fodder, pods, fruits etc. The interesting thing is that, during drought when annual crops fail, the yield of NWFPs does not decrease drastically, thus trees provide food security during drought. Besides some trees produce other products like gum, resins, edible/non-edible oils etc. which increase the income of the marginal farmers who are in the majority in this region.

The services provided by the trees in hot arid regions are equally important for moderating the climate. Cultivated in the form of shelterbelts, they reduce wind erosion and conserve soil and water.

In agroforestry systems, trees reduce the solar heat load on crops and also decrease wind speed and soil temperature, conserve soil moisture, increase biological activities in the soil and thereby improve the microenvironment for crops.

The yield of crops grown under P. cineraria is generally higher than sole crops, knowing this fact agroforestry has been followed in this region for centuries. The edapho-climatic conditions in this region are so harsh that every tree improves the environment, it may be at micro level, yet its impact is visible.

The other important NWFPs in these arid areas are plants with medicinal values. Due to climatic specialty there are medicinal plants which are either endemic to this region or their quality deteriorate if they are grow/found in higher rainfall areas.

Research Priorities: Considering the social and ecological demands, policy on agroforestry research also needs to be prioritized. Some of the possible priorities are-In-situ rainwater conservation for tree establishment and for agroforestry more generally.

Water is the most limiting factor for productivity and sustainability in the arid zone. As described earlier, the effect of bringing additional water through canals or ground water has been the cause of unsustainability in the ecosystem. Thus the only possibility remains the conservation of rainwater. Trees are the keystone of arid zone conservation and once trees are established the other conservation processes start automatically *e.g.* rainwater conservation, development of vegetation cover etc.. Further research is required on:

(i) Development of in-situ rainwater conservation techniques that require less labour and maintenance;

(ii) Development of plant-based water absorbing/retaining materials;

(iii) Initiatives for widespread adoption of rainwater conservation techniques;

(iv) Improvement in traditional water harvesting systems;

(v) Development of simplified windbreak establishment techniques.

Improvement of Policy: There are some important aspects on which improvement of policy is needed for creating a more congenial environment for development. These include:

Increased international funding for arid-zone research and development. The arid regions of the world in general suffer from a vicious cycle of low productivity, a low level of investment and as result, poverty. Greater priority is given at the international level for funding of projects and programmes related to timber forestry and rainforest management, perhaps because these activities contribute more to international trade. Very limited funds aimed at combating desertification are available to the arid lands.

Forestry and agroforestry activities in arid regions no doubt contribute less to international trade but they provide substantially support to the livelihood of people in these climatically less favored regions.

There is a need to rationalize these priorities and for the international community to give due importance to the non-tradable benefits of forestry/agroforestry, which are otherwise the basis of sustainability and livelihood. Suggested areas for increased investment in the arid zone are as follows:

(i) *Organic farming in virgin arid lands*. The philosophy of the traditional systems was to work in synergism with nature and use locally available natural resources. High input agriculture (HIA) on the other hand is only sustainable with assured input supply, *e.g.* irrigation, fertilizers etc, which is not always possible in drylands. Thus the chances of failure are much greater than in traditional systems during drought. Also due to less use of synthetic chemicals in the past, these lands are comparatively free from the residues of such chemicals. Hence organic farming, based on traditional systems and conservation of local natural resources can be one area for investment without much effort. Organic farming of NWFPs can be the first choice for investment.

(ii) *Eco-tourism*. Investment in eco-tourism can prove to be a large income generator in the arid lands. Increasing numbers of people are spending increasing amounts of time and money enjoying the unique resources of these arid land environments, *e.g.* spectacular scenic beauty, diverse wildlife population, relatively clean air and other amenities such as large open spaces. Investment in the forestry sector can be in amenity plantations, desert parks, buffer strips, roadside plantations and green belts can be economically as well as environmentally beneficial for the arid zone.

Exploring the potential of arid zone agroforestry as a tool for solving environmental problems.

Conservation of Biodiversity: Biodiversity is the essential safety mechanism of any ecosystem and as the arid zone is considered to be fragile in nature the importance of conservation of biodiversity is further increased. Because of biotic interference and developmental activities, not only is the population of various species reduced but also their natural habitats have been destroyed. Thus the easiest way to save these species is ex-situ conservation with protection. This is possible through arid zone agroforestry, in which all types of plants *e.g.* trees, shrubs, climbers etc can be accommodated, combining conservation with domestication for increased production.

Desertification Control: Desertification is the result of exploitation of arid zone ecosystems beyond their carrying capacity. The carrying capacity of the arid lands is inherently low thus they are highly prone to desertification. Through soil and water conservation and sustainable management of vegetation, the carrying capacity of natural resources can be enhanced.

Various arid zone agroforestry models, *e.g.* wind breaks, sand dune stabilization, silvi-pasture, agri-horticulture etc. can be appropriate for soil and water conservation in arid lands, and the productivity of lands suffering form desertification can be enhanced and in this way desertification can be controlled. Arid lands can be considered simulated desertified drylands for planning of desertification control.

This can be important for preventive strategies for drylands rather than to control desertification after the process starts. Any technology or strategy that is found suitable to increase the productivity of arid lands, would give better results if applied to the desertification-prone drylands because they have a better environment than the arid lands. Assessing the potential of arid zones as carbon sinks.

In addition to their other benefits, trees on farms or in pastures also store carbon. Arid lands cover almost one forth of the earth's surface and any effort to increase tree populations substantially contributes to their value as carbon sinks. Due to their perennial nature and high root turnover, grasses can also prove to be a good carbon sink.

Silvipasture systems are prevalent in most of the arid zone and when both the tree and grass components are properly managed, an increase in net carbon storage can be achieved.

The arid lands are prone to desertification and trees and shrubs can be grown on these lands to prevent further soil degradation and to produce biofuel. These trees also enhance soil organic carbon content and thereby increase soil carbon storage. About 104 million ha land area in the world is affected by strong and extreme soil erosion.

ENVIRONMENTAL SUSTAINABILITYOF AGRICULTURAL BIODIVERSITY

Each species in an agro-ecosystem is part of a web of ecological relationships connected by flows of energy and materials. Whilst each species may occupy a specific ecological niche (*e.g.* primary producer, specialist or generalist consumer, decomposer) it is involved in sustaining many different agro-ecosystem functions and environmental processes, either directly or indirectly. In this sense the different components of agricultural biodiversity are inherently multifunctional and contribute to the resilience of production systems whilst providing environmental services at the larger landscape level. However, it is important to note that some species may play a key driving role in forming the structure and overall behaviour of agro-ecosystems and landscapes at different scales. There is indeed growing evidence that the diversity and functional complexity of all ecosystems can be traced to a small number of critical structuring processes, some of which are mediated by critical "keystone species". An example is the suite of 35 species of insectivorous birds that mediate budworm outbreak dynamics in the eastern boreal forest of North America.

Farmers, herders and fishermen have often enhanced the multiple functions of agricultural biodiversity through choice of genetic material, design of cropping patterns, development of crop and livestock production systems, land and water management practices as well as institutional arrangements. Local knowledge about the properties and dynamic roles of agricultural biodiversity is crucial in this connection. For example, the *mal monte* (bad weeds) and *buen monte* (good weeds) management systems of Mexican farmers recognises that the vegetation community as a whole must be managed to promote those aspects that are beneficial.

In low external input farming, different components of agricultural biodiversity are usually combined to give practices finely tuned to the local biophysical and socioeconomic conditions of individual farmers, herders and fish culturists. Natural processes mediated by agricultural biodiversity are favoured over external inputs and by products or wastes from one component of the agro-ecosystem become inputs to another. An example is the mulberry grove-fishpond system in the Pearl River Delta of China. In this multifunctional system, the white mulberry (*Morus alba*) tree produces organic substances (mulberry leaves etc). These are used to feed silkworms that, in turn, produce their silk and chrysalides. The fallen parts of the mulberry tree and the excrement of the silkworm are applied to the fishpond where they are converted into fish biomass. The excrement of the fish, as well as other unused organic matter and bottom mud are returned to the mulberry grove as fertilizer, after being broken by a diverse suite of benthic micro-organisms. The agricultural biodiversity harnessed by the fish culturalists allows for the closing of nutrient cycles and efficient production in time and space. Fish polycultures are thus made up of species that dwell in the upper, medium and lower layers of the pond, as well a fish species with different feeding habits (*e.g.* plankton feeders, herbivorous fish, benthic mollusc feeders, and omnivorous fish).

In more specialised, high input farming based on the use of high yielding varieties, agricultural biodiversity helps sustain many production functions such as soil organic matter decomposition, pollination and pest control. In the USA or Australia for example, farmers may manage cover crops primarily to save soil and water in intensive orchard production systems. However, the species chosen will usually perform other functions in the agro-ecosystem.

In addition to protecting against soil erosion, cover crops usually enhance soil structure, improve soil fertility and nutrient cycling as well as play a role in pest management by providing habitat heterogeneity and preserving a favourable balance between pests and predators. Depending on the species, trees can also provide fodder for animals, so increasing the number of internal linkages within the agro-ecosystem. These examples highlight the multiple functions of agricultural biodiversity and are intended as a reminder that functions are discussed one by one in this report only for the sake of convenience.

Decomposition and nutrient cycling functions. The crop plants, trees, livestock and fish deliberately chosen by farmers are the main determinants of the diversity of the flora and fauna that makes up the decomposer subsystem. Available evidence shows that decomposer communities are highly diverse and are centrally involved in nutrient cycling, organic matter dynamics and other ecosystem functions. Detailed knowledge on the extent and functions of this diversity is limited; there is relatively more information on the functions of soil biodiversity than on the dynamics of decomposer communities in aquatic environments.

Some functional groups in soils are widespread in distribution (*e.g.* nitrogen fixing bacteria, mycorrhizal fungi and predators of soil borne pests) whilst other like earthworms and termites are more restricted in their distribution. A gradually emerging picture structures soil biodiversity into a series of more or less spatially independent guilds. The spatial separation between distinct guilds (surface litter, root litter, rhizosphere guilds...) allows decomposer organisms to co-exist whilst containing communities of organisms that are functionally equivalent. For instance, several ecological guilds of earthworms can be recognised in humid tropical soils with different roles in litter transformation and as "ecosystem engineers". Through their activities of feeding, burrowing and casting, they modify the physical, chemical and biological properties of soil and thus its ability to support above ground vegetation. Together with termites, different species of earthworms are among the key functional groups in humid tropical soils. At least 42 native and exotic earthworm species common in tropical agro-ecosystems have been identified as able to resist disturbances linked to agriculture and agroforestry practices and build up sizeable populations in these environments.

The sensitivity of decomposer functions to farm management practices is also evident in the mechanised agriculture of developed countries. Comparisons of the soil biological diversity in biodynamic, organic and conventional farms in Switzerland show higher species diversity and functional levels in biodynamic and organic plots than in conventional systems. The significantly higher biomass, diversity and functional activity of soil micro-organisms, earthworms, ground beetles, staphilinids and spiders found in biological systems are largely due to the organic amendments and more selective plant protection measures used in the biological systems.

In high input-high output agriculture, microbial diversity is also a central component of integrated plant nutrition systems (IPNS) that aim to maximise the efficiency of plant nutrient supplies to crops by complementing the use of on-and off-farm sources of plant nutrients. Nitrogen fixation through bacteria (notably *Rhizobia spp*) and algae (*Azolla spp*) as well as phosphorus cycling *via* mycorrhizal fungi species are particularly noteworthy in this connection. Microbial diversity is generally known to mediate nutrient cycling. However, there are few detailed studies of the dynamic role of micro-organisms in structuring landscapes and agro-ecosystems at different spatial and temporal scales.

A long term study of African savannahs has shown that the productivity of large mammalian herbivores,-upon which the human economy of the savannah is based, is dependent on water availability acting as an "on-off" switch for the mineralisation of nitrogen, phosphorus and sulphur. Some 80 per cent of the mineralisation is performed by the diverse soil micro-organisms which respond dramatically to the presence of water or rain. Where the immediate limiting resource in broad leafed savannahs is nitrogen, the key effect of water is to control the availability of inorganic nitrogen by modulating the functions of soil microbial diversity rather than control photosynthesis. The interaction between water and micro-organism activity thus sustains soil fertility. In turn, soil fertility has a profound effect on savannah ecology by determining not only plant production but also what fraction of it is edible and which species of plants and animals will be present. Livestock production is thus closely dependent on the dynamic interactions between water availability and the activity of a diverse suite of soil micro-organisms in these semi-arid landscapes.

Biomass production and yield efficiency functions. Low external input production systems usually incorporate a wide range of species and genotypes that serve a variety of production goals and are used for their resistance to diseases and pests as well as for the differential exploitation of microhabitats. The relative productivity and efficiency of these diverse agro-ecosystems (fish polycultures, mixed herds, intercrops, integrated agro-silvo-pastoral systems) have been quantified in terms of relative yield or energy efficiency of diverse units as compared with sole crops. Results indicate that diversity rich agriculture is generally highly productive in terms of its use of energy and unit land area (or unit water volume). For example, the land equivalent ratio (LER) of yields with intercrops in which mixtures include a legume species is usually significantly greater than outyield of sole crops. The energy efficiency (ratio of energy output to energy input) of pig or poultry production in internally diverse agro-ecosystems can be up to 10 times higher than that of intensive pig and poultry farms based on genetically uniform single species reared with enormous subsidies of fossil fuels. Whilst the yield output per labour hour of the more intensive and uniform systems is extremely high (in the absence of the internalisation of social and environmental costs), the more agricultural biodiversity rich systems are generally efficient producers of significant amounts of biomass. This efficiency is largely a product of the systems' biological and structural complexity that increases the variety of functional linkages and synergies between different components of agricultural biodiversity.

Soil and water conservation functions. Soil, water and nutrient conservation have been improved with the use windbreaks, contour farming with appropriate border crops and cover crops in a wide range of agro-ecosystems. In France over 150 species of trees and shrubs are used for soil and moisture conservation, with different species mixtures planted as hedges and taller windbreaks in gardens, orchards, whole farms and the larger rural landscape. In Sahelian countries of Africa windbreaks made up of *Euphorbia tirucali*, *Parkinsonia aculeata, Opuntia tuna* and *Prosopis africana* trees and shrubs help to conserve soil and moisture, and raise the yields of cereals which are grown between. In Mexico, contour lines are often planted with *Agave americana* to conserve soils and retain moisture whilst in southern Italy and Greece *Opuntia tuna* performs similar functions. Many of these plants are multiple purpose species yielding wood, edible fruit and nuts, fodder, refuges for natural enemies of pests, nitrogen biofixation and medicines in addition to their soil and water conservation functions. Cover crops consist of plant species that are deliberately established after or intercropped with a main crop to serve various regenerative and conservation functions including soil and water conservation. Annual bluegrass, lana vetch, crimson clover, black medic, purple vetch and barley are some of the species recommended as cover crops for orchards and vineyards in California, USA. The wide variety of management systems in high input-high output orchards and vineyards creates a demand for a diversity of cover crops. Grass species have fibrous root systems that make them particularly useful in building soil structure, providing erosion control, and improving water penetration. Legumes are not as effective as grasses in improving water penetration but they contribute nitrogen to the soil. Many cover crop options can be selected for soil and water conservation from a diversity of annually seeded winter growing grasses and legumes, summer annuals, perennial grasses and legumes and reseeding winter annual grasses and legumes.

In North America and Europe, living mulch systems can be an economic way for commercial soybean, corn and vegetable growers to reduce soil erosion and water loss, increase soil organic matter and keep yields constant in high input-high output agroecosystems. Legume species commonly used as living mulches include alfalfa, short white clover, hairy vetch and red clover.

Pest control functions. Agricultural biodiversity in the form of predators, parasitic wasps, microorganisms plays a key role in controlling agricultural pests and diseases. For example, according to CAST (1999) more than 90 per cent of potential crop insect pests are controlled by natural enemies that live in natural and semi-natural areas adjacent to farmlands. They have estimated the substitution of pesticides for natural pest control services at a cost of $54 billion per year.

Many methods of pest control,-both traditional and modern-, rely on biological diversity. The development of crop varieties and animal breeds that are resistant to specific pests and diseases selectively draws on the genetic diversity available *in situ* and in *ex situ* collections of germplasm.

Genetic mixtures deployed in temperate and tropical agroecosystems can be effective in containing diseases in small grain crops as well as insect outbreaks in cassava, corn and potato for example. There are also many documented experiences showing that insect pests tend to be less abundant and damaging in agroecosystems with higher plant diversity *e.g.* intercrops, polycultures, crop rotations, cover crops, mixed tree stands, mixtures of annual and perennial plants.

Depending on the pest species and the context, the plant diversity acts to reduce pest damage by interfering with the host seeking and reproductive behaviour of the pest, by enhancing the pests' natural enemy populations or by a combination of these processes. Judicious vegetation management within and around agroecosystems can thus enhance biological control or confer an overall resistance to pests and disease outbreaks.

Understanding how agricultural biodiversity directly or indirectly affects pest and disease dynamics is critical for the design of pest management at different scales. For instance, recent work in Javanese rice fields shows that there is an enormous diversity of arthropods, even in high input-high output agriculture. The arthropod communities are structured such that the dynamics of seasonal succession consistently lead to high levels of pest suppression, with little chance of outbreak. From the time that water first floods a farmer's field in preparation for planting, organic matter,-derived from residues from the previous crop cycle, organic waste in irrigation water, and algal growth-, provides the energy for an array of microorganisms (bacteria and phytoplankton) and detritus-eating insects. The adults of the plankton-feeders (midges and mosquitoes) together with the detritus-feeders, provide a consistent and abundant source of alternative food for generalist predators very early in the season.

As a result, pest mortality due to predation is high from the very earliest part of the season; hence, minimising the chance of damaging pest outbreaks. However, this intrinsic strength and stability of the rice agro-ecosystem is influenced by two main, large scale, external factors: 1) local and regional patterns of pesticide use, and 2) landscape effects-specifically, the spatial scale at which fields are synchronously planted, the duration and nature of fallow periods, degree of surrounding weedy or natural vegetation and existence of nearby ponds or other sanctuaries for natural enemies. This type of information on the functions of agricultural biodiversity in rice paddies provides the ecological basis for integrated pest control through careful management of the wider landscape and decisions on pesticide use.

Pollination and dispersal functions. There are more than 100,000 known pollinators (bees, butterflies, beetles, birds, flies, and bats). Pollination mediated by components of agricultural biodiversity is an important function in a variety of terrestrial agroecosystems (biotic pollination *per se* is poorly represented in aquatic ones). About half of all plant species, including food-producing crop species, are pollinated by animals. For example, the pollination of various fruit crops by bees and other insects is critical in mountain areas of Asia.

In Nepal *Apis cerana* begins foraging at temperatures 5-7°C lower than those that initiate *Apis mellifera* foraging. Managed crop pollination with a variety of bee species in different zones plays an important role in overall agricultural development. The benefits of pollination are also considerable in high input-high output agriculture: the economic value of pollination services in the United States is estimated in billions of dollars per year. Management practices that reduce the species or abundance of pollinators can result in less genetic variation in crops dependent on pollinator visits for reproduction, both in temperate and tropical agriculture. With a loss in pollinators, seed production declines and the vulnerability to pests and climatic change increases with the resulting loss of genetic diversity.

"Mobile link" species (*i.e.* animals necessary for the persistence of plant species that in turn support otherwise separate food webs) such as pollinators and seed dispersal agents may be critical to the maintenance of the species richness of tropical forest based agroecosystems and complex home gardens imitating the natural forests' architecture. Many species in tropical forests managed for food and agriculture depend on a small suite of frugivores for dispersing their seeds. Loss of these species of fruit eaters may adversely affect the long-term viability of many tree species important for food security. Reductions in genetic variability are likely to be high for plant species that are highly dependent on frugivory for seed dispersal.

Biological diversity conservation functions. There is no strict divide between "wild" and "domesticated" species important for food and livelihoods. Many wild plant species and populations that have been considered to be wild are in fact carefully nurtured by people. A similar continuum exists for animal species that use agroecosystems as habitat, nesting grounds and food. Whilst not necessarily the subject of conscious management by herders or farmers, many wild species thrive in, or are dependent on, agroecosystems. In general the more structurally and biologically complex the agroecosystems, the more diverse the forms of wildlife. For example:

- The systems of shifting cultivation common in Asia have long been considered to be damaging to biological diversity in particular. However, given the opportunity, shifting cultivators preserve wild resources. Studies in Asia have concluded that most of the mature forests in this region are not virgin forests, but merely old forests that have reached a relatively stable dynamic state after some earlier clearing by human or natural means. Under traditional systems of shifting cultivation which speed up natural succession, wildlife flourishes, with elephants, wild cattle, deer, and wild pigs all feeding in the abandoned fields. Tigers, leopards and other predators are in turn attracted by the herbivores. Older fields contain a high proportion of fruit trees that attract primates, hornbills, squirrels and a variety of other animals.
- Long term research in the Lowland of west Poland shows that a mosaic of landscape structure of small cultivated fields, shelterbelts, meadows and small ponds helps maintain biological diversity (and many other functions such as enhanced water storage). Mammal, amphibian, reptile, insect diversity are enhanced and the number of

bird species is positively correlated with the diversity index of the landscape.

Climate functions. Although agricultural and the atmosphere interact in many ways, the links between climate and weather, and agricultural biodiversity can be rather complex.

At the global scale, the distribution of individual plants and animals, vegetation and crops is conditioned by available climatic resources such as solar radiation, which controls the production potential, and by rainfall, which determines to what extent the radiant energy can actually be used by plants for their growth. Indeed, it is for these reasons that climatic classifications largely coincide with vegetation maps. At the local scale, types of landscapes and vegetation, including crop and crop-vegetation mixes contribute towards modifying the local climates by directly affecting wind patterns, rates of evaporation, rainfall interception (effective rainfall), etc. This in turn conditions the development of vegetation and crop canopies which can be said to create their own microclimate, resulting from the interaction of plants with the general climate. Shelter belts, or the use or large tree belts to protect tropical plantation crops from cyclones, provide clear examples in which people and communities derive direct advantage from these landscape features.

Many examples demonstrate the benefits of maintaining minimum agrobiodiversity in the face of climate variability. Recent droughts in some southern African countries for example, have shown that mixes of local varieties planted over several weeks have the potential to better resist unusual patterns of rainfall variability (*e.g.* early-season drought) than some modern varieties planted on the "optimum dates". In fact, there appears to be a link between crop biodiversity and relatively low but regular production, one of the keys to food security.

Similarly, atmospheric pollution has been shown to interact with agro-ecosystem functions, both positively (nutrients, such as nitrates and sulphur) and negatively (toxic compound like tropospheric ozone and heavy metals). Given that the response to these substances varies enormously across the spectrum of species, they have the potential to modify patterns of biological diversity.

Agriculture also contributes towards the emission of some of the most significant "greenhouse" gases (in terms of global warming potential, GWP). Some studies have suggested for example, that increases in the area of permanently or quasi-permanently flooded rice cultivation account for a significant proportion of the increase in net methane emissions. The increased specialisation of ruminant production based on high yielding breeds has also significantly contributed to global methane emissions (methane is produced by anaerobic digestion in animals). An extreme case would be the enhanced biogenic emissions of carbonylsulphide associated with increasing cultivation of high sulphur crops such as rape. Oxidation of carbonylsulphide in the stratosphere leads to the production of sulphate aerosols that influence the intensity of ultra violet radiation reaching the earth's surface, potentially affecting the dynamic functions of agricultural biodiversity. Volatile organic compounds such as terpenes and isoprenes are produced and released into the atmosphere by many plant species, especially in agroecosystems dominated by Mediterranean shrubs, eucalyptus and conifers. By influencing the oxidation capacity of the troposphere, these volatile compounds influence the abundance and distribution of trace gases such as ozone. In almost all cases however, many of the these effects are marginal when compared with the climatic effects of land use change, including deforestation, which have until recently constituted one of the main non-industrial sources of carbon dioxide

Functions in the water cycle. There are few experimental studies exploring the links between agricultural biodiversity and water. However, available evidence shows that agricultural biodiversity plays a crucial role in cycling water from the soil to the atmosphere and back. It also has measurable impacts on water quality. Agroecosystems with different species assemblages and plant architectures result in differences in the amount of precipitation intercepted, the proportion of precipitation converted to stem flow, and the proportion of precipitation that infiltrates the soil rather than running off. At the landscape level, the types, relative abundances and relative spatial locations of agro-ecosystem types affect the amount of water moving from one point to another. For example, conversion of vegetation within a watershed from forest or shrub land to less structurally and biologically complex grassland is known to influence stream flow out of the watershed, in both temperate and tropical systems.

At the functional group level, the root structure, phenology and physiology of different plant species important for food and agriculture have direct implications for the quantity and timing of water transfer to the atmosphere *via* evapotranspiration. Individual plant species differ in their resistance to water stress, their efficiency of water uptake from the soil, and so on. Genetic variations among crop varieties (differences in water use efficiency, stomatal resistance...) also influence these processes. At the landscape level, evapotranspiration from agroecosystems can have an effect on relative humidity and microclimate downwind.

In both terrestrial and aquatic environments, the plant and animal components of agricultural biodiversity also function to alter water quality by performing various filtration, uptake and excretory processes. These functions all affect the composition and concentration of dissolved gases, solutes and particulates. Species level differences in physiology can positively or negatively affect water quality. However, in most cases a greater diversity of biological organisms (from microbes to fish and macrophytes) leads to a higher quality of water for human consumption and use.

The influence of agricultural biodiversity on landscape structure. A landscape is a heterogeneous area made up of a cluster of interacting ecosystems that is repeated in similar form throughout. The spatial layout between landscape elements together with the interactions and linkages between them determine the landscape's structure and functions. The many different species found in a landscape are essential components of that landscape. By providing environmental services and functions agricultural biodiversity can have a profound influence on landscape structure. Through its positive or negative effects on agricultural biodiversity, human activity can transform whole landscapes over large areas.

For example, many rural communities enrich their agricultural plots and forest fallows with valued perennial plants. Through such enrichment practices, successional vegetation can become a site for economic production as well as for ecological rehabilitation. Each of the major tropical forest regions has many economic woody plants that have been managed, probably for millennia, in enriched fallows. In Vanuatu the natural composition of forests has been dramatically altered by centuries of itinerant gardening, favouring tree species that bear edible fruits and nuts. Fallows have been (and still are) enriched with rattan in East Asia, rubber in Sumatra, *Casuarina* in Papua new Guinea, *Gliricidia* and peach palm in Central America, oil palm in West Africa, and edible fruits and nuts universally. Locally adapted enrichments have altered species composition and also directly influenced the structure of landscapes at different spatial scales.

The influence of agricultural biodiversity on landscape structure is partly determined by the social institutions that mediate the relationships between rural people and the environment, and partly by climate and edaphic factors. For example in the semi-arid landscapes used by pastoralists of Africa, there are high levels of spatial and temporal variability in fodder biomass production, highly variable rainfall and episodic chance events such as drought. In these non-equilibrium systems pastoralists have developed opportunistic management schemes to exploit the patchiness of the vegetation, learnt to avoid risks by moving herds and flocks to make best use of heterogeneous landscapes and diversified their livelihood activities.

Pastoralists have rules and regulations which govern the use of water, pasture, animal movement and control of vegetation and trees. Management of agricultural biodiversity and the larger landscape is mediated by these local institutions. The local adaptive management of agricultural biodiversity enables people to cope with uncertainty and sustain the dynamic environmental processes that define and shape those landscapes. This is in stark contrast with the degradation that occurs under the centrally planned, standardised rangeland and livestock management schemes often based on erroneous concepts of carrying capacity and equilibrium ecology.

New perspectives in ecology have challenged the conventional views of drylands in Africa as stable ecosystems subject to decline and desertification once carrying capacity is exceeded. Rangelands and pastoral landscapes are resilient and less prone to degradation and desertification than once thought. The new findings concord with the knowledge of many local livestock herders and emphasise how rangelands are subject to high degrees of uncertainty and ecological dynamics, characterised by sudden transitions rather than slow and predictable change.

Specific components of agricultural biodiversity are often directly implicated in the processes that structure agroecosystems at different temporal and geographical scales (from small farm plots to whole water/landscapes). Even highly complex landscapes like tropical irrigated rice or forests in the savannah transition zone of West Africa, are apparently structured by a very few key variables. Research over the past 20 years in applied ecology of managed systems shows that ecosystem and landscape dynamics tend to be organised around a small number of nested cycles, each driven by a few dominant variables.

A small number of plant, animal, and abiotic processes structure biomes over scales from days and centimetres to millennia and thousands of kilometres. Individual plant and biogeochemical processes dominate at fine, fast scales; animal and abiotic processes of mesoscale disturbance dominate at intermediate scales; and geomorphological ones dominate at coarse, slow scales....the physical architecture and the speed of variables are organised into distinct clusters, each of which is controlled by one small set of structuring processes. These processes organise behaviour as a nested hierarchy of cycles of slow production and growth alternating with fast disturbance and renewal.

Identifying and understanding the dynamics of these "structuring variables" provides a practical basis for sustainable agriculture and landscape management.

8

Global Discourse on Forestry

"Liberty and forest laws are incompatible," remarked an English country vicar, speaking on behalf of villagers shut out of woodland reserved for the exclusive use of the king, in 1720. The history of state forestry is indeed a history of social conflict. In monarchies and in democracies, in metropolitan Europe as well as in colonial South Asia, the state management of forests has met bitter and continuous opposition. On the one side are the professional foresters who believe that timber production can be ensured only through the exclusion of humans and their animals from wooded areas; on the other, the peasants, pastoralists, charcoal ironmakers, basketweavers, and other such groups for whom access to forests and forest resources is crucial to economic survival. Environmentalists have added to the criticisms of these latter groups, charging foresters with simplifying complex ecosytems in the direction of commercially valuable but biologically impoverished monocultures.

These contending parties have battled for more than two hundred years. In continental Europe, the eighteenth and nineteenth centuries were peppered with social protest movements against the state management of forests. These protests inspired, among other things, Karl Marx's first political writings and a memorable novel by Honore de Balzac capturing peasant hostility to forest officials. When the European model of strict state control over forests was exported to the colonies, the disaffected peasants and tribals responded with arson and violence. Movements over forest rights were a recurring phenomena in colonies ruled by the British, the Dutch, and the French. The conflicts persisted when the post-colonial governments of countries such as Malaysia and Indonesia followed the authoritarian model of forest management inherited from the colonizer.

In recent decades, however, the global discourse on forestry has moved towards a more accommodationist perspective. Foresters and peasant protesters now seem to talk to, rather than talk past, each other. A willingness to listen to and at least partially incorporate the other point of view has replaced the rigid and uncompromising attitude of the past. Within the forestry profession itself, skeptics doubt the contemporary relevance of the custodial and policing approaches previously followed. A system of natural resource management crafted in absolutist and colonialist times clearly needed to be seriously modified or even overthrown. Social activists and community leaders have, meanwhile, moved from demanding a total state withdrawal from forest areas to asking governments to more seriously and sympathetically consider the rights of forest-dependent communities.

With this move from conflict to collaboration have come shifts in the language of forestry itself. Terms such as "scientific forestry" and "rational land management," euphemisms for state control and commercial timber production, are being rapidly replaced by sweet-sounding phrases such as "community management," "participatory development," and "joint forest management." While these terms have come into vogue in the last two decades, they have, in fact, a very long geneaology. From the beginnings of state forestry, there have been serious attempts to democratize the regimes of resource management. Both dissidents within the bureaucracy as well as intellectual activists outside it tried hard to make the state respond more sensitively to the just claims of local communities. The ongoing programmes of joint forest management in India can draw legitimacy and sustenance from a struggle that is at least a century old.

THE LAW AND THE PROTESTS

The crucial watershed in the history of Indian forestry is undoubtedly the building of the railway network. In a famous minute of 1853, the governor general of India, Lord Dalhousie, wrote of how railway construction was both the means for creating a market for British goods and the outlet for British capital seeking profitable avenues for investment. Thus between 1853 and 1910 more than eighty thousand kilometers of track were laid in the subcontinent.6 The early years of railway expansion witnessed a savage assault on the forests of India. Great chunks of forest were destroyed to meet the demand for railway sleepers (over a million of which were required annually). The sal forests of Garhwal and Kumaun, for example, were "felled in even to desolation." "Thousands of trees were felled which were never removed, nor was their removal possible."

This depradation brought home most forcefully the fact that India's forests were not inexhaustible. At this time the British were unquestionably the world leaders in deforestation, having burnt or felled hundreds of thousands of acres of woodland in Australia, southern Africa, northeastern United States, Burma, and India. Knowing little of methods of sustained-yield forestry, they called in the Germans, who did. Thus in 1864 they established the Indian Forest Department, which for the first twenty-five years of its existence was serenely guided by three German inspectors general of forest-Dietrich Brandis, Wilhelm Schlich, and Bertold von Ribbentrop. For its effective functioning, the new department required a progressive curtailment of the previously untrammeled rights of use exercised by rural communities all over South Asia. An act was hurriedly drafted to establish the claims of the state to the forestland it immediately required, subject to the provision that existing rights not be abridged. This act was "infinitely milder and less stringent than that which is in force in most European countries." The search commenced for a more stringent and inclusive piece of legislation. In 1869, the Government of India circulated to the provinces a new draft act, which sought to strengthen the state's control over forest areas through the regulation and in some cases extinction of customary rights.

The new legislation was based on the assumption that all land not actually under cultivation belonged to the state. Of course, it was not easy to wish away the access to forests exercised in centuries past by peasants and other rural groups. The colonial state, however, argued that such use, however widespread and enduring, had been exercised only at the mercy of the monarch. Unless it had been expressly recorded in writing, customary use was deemed to be a "privilege," not a "right." And since the British government was the successor to Indian rulers, the ownership of forests and waste was now vested in it. "The right of conquest is the strongest of all rights," emphatically remarked one forest official. "It is a right against which there is no appeal."

There were, however, some notable dissenting voices within the colonial government. Sent the draft forest bill by the Government of India, the Madras Government in turn invited responses from various officers. The views of Narain Row, Deputy Collector of Nellore, are representative. The proposed legislation, he said, had no historical precedent, for "there were originally no Government forests in this country. Forests have always been of natural growth here; and so they have been enjoyed by the people."

Another Deputy Collector, Venkatachellum Puntulu, of Bellary, argued that the burden of the new legislation would fall most heavily on the poor. While large landlords would find it relatively easy to deny the state any claim over their forest property, unlettered peasants would not be able to prove rights of ownership, even though they traditionally used forests as common property.

Criticizing the detailed rules prohibiting the collection of different kinds of forest produce, Puntulu penetratingly remarked that "the provisions of this bill infringe the rights of poor people who live by daily labour (cutting wood, catching fish and eggs of birds) and whose feelings cannot be known to those whose opinions will be required on this bill and who cannot assert their claims, like [the] influential class, who can assert their claims in all ways open to them and spread agitation in the newspapers."

After several such responses came in, the Madras Board of Revenue told the Government of India that the claim of the state to uncultivated forests and wastes was virtually nonexistent:

There is scarcely a forest in the whole of the Presidency of Madras which is not within the limits of some village and there is not one in which so far as the Board can ascertain, the state asserted any rights of property unless royalties in teak, sandalwood, cardamoms and the like can be considered as such, until very recently. All of them, without exception are subject to tribal or communal rights which have existed from time immemorial and which are as difficult to define as they are necessary to the rural population.... [In Madras] the forests are, and always have been common property, no restriction except that of taxes, like the Moturpha [tax on tools] and Pulari [grazing tax] was ever imposed on the people till the Forest Department was created, and such taxes no more indicate that the forest belongs to the state than the collection of assessment shows that the private holdings in Malabar, Canara and the Ryotwari districts belong to it.

The Madras Government advanced three basic reasons for rejecting the bill drafted by the Government of India: First, because its principles, scope and purpose are inconsistent with the existing facts of forest property and its history. Second, because, even if the Bill were consistent with facts, its provisions are too arbitrary, setting the laws of property at open defiance, and leaving the determination of forest rights to a Department which, in this Presidency at all events, has always shown itself eager to destroy all forest rights but those of Government. Third, because a Forest Bill, which aims at the regulation of local usages ought to be framed.

The objections were disregarded, and in 1878, the new bill passed. The act divided the forests of the subcontinent into three broad classes. State or reserved forests were to be carefully chosen, in large and compact areas that could lend themselves to commercial exploitation.

The constitution of these reserves was to be preceded by a legal settlement that either extinguished customary rights of user, transferred them as "privileges" to be exercised elsewhere, or, in exceptional cases, allowed their limited exercise. In the second class, of "protected" forests, rights and privileges were recorded but not settled. However, all valuable tree species were to be declared as "reserved" by the state, while the Forest Department had the power to prohibit grazing and other ostensibly damaging practices.

The Forest Act also provided for a third class of forests-village forests. But as these lands had first to be constituted as reserved forests, the procedure aroused suspicion among the villagers, and this chapter remained a "dead letter."15 Meanwhile, the area of forests under strict state control steadily expanded. In 1878, there were 14,000 square miles of state forest. By 1890, this had increased to 76,000 square miles, three-fourths of which were reserved forests. Ten years later, there were 81,400 square miles of reserved forests and 8,300 of protected forests. Given increasing demand for wood products, the state sought to establish firmer control over forests, both by expanding the area taken over under the Forest Act and by converting protected forests to reserved forest. The Indian Forest Act of 1878 was a comprehensive piece of legislation that came to serve as a model for other British colonies.16 Within India, it allowed the state to expand the commercial exploitation of the forest while putting curbs on local use for subsistence. This denial of village forest rights provoked countrywide protest.

REBELLIONS AGAINST COLONIAL FORESTRY

The history of colonial rule is punctuated by major rebellions against colonial forestry-in Chotanagpur in 1893, in Bastar in 1910, in Gudem-Rampa in 1879-80 and again in 1922-23, in Midnapur in 1920, and in Adilabad in 1940. These rebellions sometimes extended over several hundred square miles of territory, involved thousands of villagers, and had to be put down by armed force. Even where discontent did not manifest itself in open rebellion, it was expressed through arson, noncompliance, and breaches of the forest law.

The participants in these protests were unlettered peasants and tribals, and we know far more of their deeds than their words.

Nonetheless, their voices do figure here and there in the archives of the state, sometimes mediated by the language of the officials reporting them. Thus in the 1880s, when the government of the Bombay Presidency was aggressively demarcating the rich teak forests of the Dang district, preparatory to their constitution as state reserves, a Bhil tribal chief sent in a petition stating that "we do not wish to let the Dang jungle [be] demarcated, for thereby we shall lose our rights and we and our poor rayat [cultivators] shall always be under the control of the Forest Department and the Department will always oppress us."

Around the same time, the colonial state was attempting to take over the deodar (cedar) forests of the upper Jamuna valley. These trees had suddenly become market-worthy, to service the then expanding railway network. But as a peasant bitterly observed, "the forests have belonged to us from time immemorial, our ancestors planted them and have protected them; now that they have become of value, government steps in and robs us of them."18 Or consider, finally, these remarks of an administrator in the Bastar district of central India, on the determination of his tribal subjects to continue practicing swidden cultivation in what was now "government" forests: "On the road from Tetam to Katekalyan I found general dissatisfaction at the restriction of penda [swidden] cultivation. I was unable to convince them of its evils [sic]. Podiyami Bandi Peda of Tumakpal has to get his son married and for this purpose he wants to cultivate penda in the prohibited area. I told him he should not do it. He replied plainly that he would cultivate it and go to jail as he had to get his son married."

In 1871, the Madras Government predicted that the new act, if passed into law, would "place in antagonism to Government every class whose support is desired and essential to the object in view, from the Zamindar [landlord] to the Hill Toda or Korombar." This was an astonishingly accurate prediction, for the Forest Department was unquestionably the most unpopular arm of the British Raj. The story of the numerous popular movements against state forestry, so long neglected by historians, is now attracting an array of chroniclers. The critics were principally of two kinds. On the one side were scholars and politicians with a deep knowledge of rural conditions, and who sometimes formed part of popular movements themselves.

Their criticisms of state forestry thus drew richly upon the feelings and grievances of the people most affected by it. On the other side were the rare (but, for that reason, significant) dissidents within the colonial bureaucracy, who opposed the centralizing thrust of government forest policy. The first set were outsiders so far as the apparatus of rule was concerned, but insiders with respect to popular opinion and popular consciousness. The second set were by virtue of race and status outsiders to Indian society, but insiders with regard to the policy of the state and the functioning of government.

PRECOCIOUS PROPHETS

In 1878, the Poona Sarvajanik Sabha, a vastly respected nationalist organization in western India, bitterly opposed the new Forest Act. Despite its middle-class origins, the Sabha had consistenly fought for the rights of the cultivator, urging that the colonial government lessen its burden of taxation on the peasantry.21 Now, in the context of the Forest Act debate, it pointed out that state usurpation was grossly violative of customary rights over forests, for both "private grantees and village and tribal communities" had "cherished and maintained these rights with the same tenacity with which private property in land is maintained elsewhere." The Sarvajanik Sabha did not, however, merely oppose the proposed Forest Act for its excessive emphasis on state control; it offered a more constructive and creative alternative.

Thus the Sabha argued that the better maintenance of forest cover could more easily be brought about by taking the Indian villager into confidence of the Indian Government. If the villagers were rewarded and commended for conserving their patches of forestlands, or for making plantations on the same, instead of ejecting them from the forestland that they possess, or in which they are interested, emulation might be evoked between neighboring villages. Thus more effective conservation and development of forests in India might be secured, and when the villagers have their own patches of forest to attend to, government forests might not be molested. Thus the interests of the villagers as well as the government can be secured without causing any unnecessary irritation in the minds of the masses of the Indian population.

The Sabha was advocating a far more democratic structure of forest management than that envisaged by the colonial government. Indeed, it was proposing the institution of a Vrikshamitra (Friends of the Trees) Award, one hundred and ten years before the Indian Government's Ministry of Environment and Forests conceived and named such a scheme, for rewarding individuals and communities who had successfully protected or replenished forest areas. Three years after the 1878 act was passed, the impact of state forestry on rural communities was foregrounded by the social reformer Jotirau Phule. Phule himself was a gardener by caste, and in general exceptionally alert to the problems of the agricultural classes. The following is his description of the impact of the Forest Department on the livelihood of farmers and pastoralists in the Deccan countryside: In the olden days small landholders who could not subsist on cultivation alone used to eat wild fruits like figs and jamun and sell the leaves and flowers of the flame of the forest and the mahua tree [both common trees of the Indian forest].

They could also depend on the village ground to maintain one or two cows and two or four goats, thereby living happily in their own ancestral villages. However, the cunning European employees of our motherly government have used their foreign brains to erect a great superstructure called the forest department. With all the hills and undulating areas as also the fallow lands and grazing grounds brought under the control of the forest department, the livestock of the poor farmers do not even have place to breathe anywhere on the surface of the earth.

THE LIMITS OF GLOBAL SUSTAINABLE MANAGEMENT

Although steps have been taken between Rio 1992 and New York 1997 in the direction of the globalization of environmental protection, the world seems to be further away from sustainability today than it was then. With environmental globalization and the consensual concept of sustainable development, the perception of an ecological crisis, or at least of ecological limits to development, appears to have vanished: acting for environmental protection increasingly tends to be seen as a technical problem, the task of increasing efficiency and of better using resources such as science, technology, information, capital, and institutions. The cause of environmental problems is no longer perceived as linked to industrial development and ever-increasing material accumulation, but it has become the very existence of human beings. Environmental problems are understood as unavoidable, as side effects of human activities, and efforts are then directed at solving these problems.

Indeed, global environmental management and sustainable development can be seen as "problem solving" concepts, in Robert Cox's terminology, as they only represent a strategy to allow the pursuit of present lifestyles and standards.

Following Strange's call for a critical International Political Economy and the need to address the question of "who gets what, how and why," the analysis of the evolution of the system of global environmental management has revealed that it tends to strengthen the mechanisms of exclusion and inequality. Global environmental management and sustainable development tend to be uneven concepts, as they do not aim at promoting the correction of global disparities. They attempt to offer a universal framework in which the global society is the unit of analysis and a large share of the blame for environmental degradation rests on the Third World. Instead of stressing affluence, over-production and over-consumption in advanced countries as the main causes of environmental degradation, it tends to suggest that problems arise from poverty. Environmental degradation is transformed from a problem of affluence into a problem of poverty. The responsibility is shifted from major polluters and industrialized countries' abuses to all inhabitants of the planet. For Chatterjee and Finger (1994), the only different element in this approach is that development is now looked at from a global perspective, making the development discourse universal.

And the New York Summit represented a step further in that direction, asserting the desirability of the globalization process and underlining its beneficial aspects. It also consolidated the role of business and industry as privileged partners of the United Nations, establishing permanent contact and consultation on environmental issues. The regulatory situation relating to TNCs and business in general has worsened greatly in the past five years.

Already in 1992, the US government successfully pressured for downsizing the UN Centre on Transnational Corporations (UNCTC), which had been set up to monitor the social, economic and environmental impacts of corporate investment in developing countries. Today, the UN is considering cutting a sub-group of the UN Human Rights Commission which addresses the impacts of corporations on a broad spectrum on rights issues. The main international initiatives and institution for establishing guidelines for the behavior of corporations, which could together lay down a code of obligations and rights of TNCs and states, have disappeared.

In their place has come a strong and growing opposite trend to reduce and remove regulations that governments have over corporations, to grant them increased rights and powers, and to reduce the authority of states to control their behavior and operations.

The Uruguay Round for example has already granted far higher standards of intellectual property rights protection to corporations, thus facilitating further their global monopolization of technology and ability to make profit through higher prices. There are also strong pressures from Northern governments at the World Trade Organization to grant foreign companies the right of entry, establishment and national treatment in all WTO member states.

In addition, the partnership between the UN and global corporations seems to have been further strengthened in the past two years. In part due to the difficult financial and political situation in which the UN finds itself as a result of the US government's refusal to pay the US $1.6 billion it owes, the UN is now openly seeking political and economic support from corporations. At the last Davos Economic Forum in Switzerland, UN Secretary General Kofi Annan called for a human face to the global market and challenged business leaders to adhere to universal values defined by the UN and contribute to global environmental protection, indicating a broader trend of growing UN collaboration with transnational corporations.

Recently, the United Nations Development Programme (UNDP) has solicited funds from global corporations with poor records on human rights, labour and the environment, such as Dow Chemical, Citibank and Rio Tinto Plc, in exchange of special UNDP sanctioned logos for use by corporate sponsors. Called the "Global Sustainable Development Facility (GSDF)," the plan calls for corporate sponsors to funnel donations to a separate entity which they will manage. In the words of the UNDP, the GSDF "brings together leading global corporations and the UNDP, to jointly define and implement a new facility to eradicate poverty, create sustainable economic growth and allow the private sector to prosper through the inclusion of two billion new people in the global market economy."

According to the internal memo, sponsors will benefit from the advice and support of UNDP through a special relationship, allowing corporations unprecedented access to UNDP's network of offices, high level governmental contacts and the knock-on effects of its reputation. The plan, revealed through a leaked internal UNDP memo, has been heavily criticized by observers and NGOs, who warn that the interests of global corporations are often at odds with the basic economic and social needs of the world's poor and the values of human rights and environmental protection the UN is meant to protect.

According to Ward Morehouse, President of the US Council on International and Public Affairs, the UN should be monitoring the human rights and environmental impacts of corporations in developing and industrialized nations, not granting special favors... Increasing collaboration will lead to a reluctance to criticize corporations which are central players in the human rights, environmental and developmental dramas unfolding every day across the globe.

To conclude, the UN and international organizations in general seem to be moving towards the adoption of a market-oriented global model of environmental governance, which sees economic globalization as a positive and integrative process. The key actors of economic globalization, transnational corporations, are taking the leading role and consolidating their influence on the system of international environmental governance. With the adoption of this project of global environmental management, one particular understanding of the world, the one promoted by business and large corporations in Western affluent societies, becomes hegemonic and appears to be universal.

Environmental concerns have been incorporated as a mere dimension of the "globalization project," understood in McMichael's definition as "an emerging vision of the world and its resources as a globally organized and managed free trade/free enterprise economy pursued by a largely unaccountable political and economic elite". This project advocates a universalized model of production, of consumption, and thus of dealing with problems of environmental protection resulting from these activities.

By assuming its universality, it tends to marginalize other knowledge and other solutions to problems of environmental protection. Interestingly enough, it was unanimously recognized that the most positive result and follow-up of UNCED was without doubt realized at the micro-level, within the Local Agenda 21 framework. In an effort to implement Agenda 21 locally, social groups have worked together with local authorities to make sustainable development a reality at the local level, often on a truly participatory basis and reflecting grassroots concern and involvement.

From a critical point of view, global management's failure, exemplified at the New York Summit, was not entirely unexpected. Indeed, the global management approach inspired by business perspectives and propagated by the international development establishment tends to strengthen the globalization process at work today, failing to counter its effects in terms of social exclusion and environmental destruction. It tends to weaken social protection and environmental protection in the name of economic efficiency.

It stands at odds with the commitment to social change and to equity that lies at the root of a critical, political economy view of global environmental politics as inserted within the dynamics of economic accumulation and social structures. The question of the ownership of natural resources, for example, is not addressed. However, environmental problems in the South are often linked to problems of resource ownership and equity.

Sustainable Development as defined in Rio and reasserted in New York has been practically translated into technocratic responses to what are in reality political problems.

The purpose of this article was to analyze and evaluate the process of "mainstreaming" of environmental concerns. Recalling the radical and transformative origins of the ecological project, it has provided evidence that environmental concerns have been remodeled by the joint action of technocratic environmentalists, the international UN-related development establishment and business and industry sectors.

Examining the results of international cooperation, the article has questioned the nature of the sustainable development consensus, a consensus deeply marked by the growing access and influence of global corporations on UN activities. Today, market-oriented perspectives to environmental problems seem to be prevailing over more transformative views, especially at the international level, within the framework of international organizations and institutional agreements.

The article has suggested that the mainstream approach to sustainable development tends to reduce ecology to a set of managerial practices aiming at resource efficiency and risk management. In doing so, it tends to address a civilizational impasse as a mere technical problem. The mainstream approach proposes that environmentalists should operate using the language and the worldview of Western economics in approaching ecological concerns.

Instead of designing cultural and political limits to development, the project of "global environmental management" tends to become part of a technocratic effort to sustain industrial development in the age of economic globalization. Environmental protection, together with democracy, human rights and free market economics, becomes a universal consensus, a universal consensus which, as Baudrillard remarks, arouses suspicion, since it is about values that have become devalued, values becoming emptied at the very moment of their hegemony. Environmental concerns become just another element in a process leading to global uniformity, a uniformity of cultures, lifestyles, mentalities, but also of relationships with nature.

This market-oriented agenda may provide a starting point for dealing with global environmental problems.

The documents, which emerged from international environmental negotiations from Rio to New York, replete with inconsistencies, represent a complex mix of disagreements, hopes and compromises. One may concede that conventions and obligations reflect the need for government negotiators to find the minimum agreeable grounds to initiate a large open-ended process on major environmental issues.

In this sense, they only produced a general framework for negotiations, steps on the way to building international regimes. They do not form a series of real commitments representing an effective consensus on how to deal with global environmental issues.

Yet considering the amount of time, energy and resources invested in this process of international environmental regime building, one might have hoped for more concrete, positive results. The failure of the present framework to effectively promote sustainability, which became evident in the 1997 New York summit, is recognized even by one of the major promoters of this path, Maurice Strong, ex-secretary general of UNCED, today President of the Earth Council. For Strong, unfortunately, the economic, social and demographical forces that lead to unsustainable development still prevail. Strong sees the lack of political will from governments as the main cause of this failure.

The present framework appears to contain many contradictions that limit the ability of the international and national communities to solve satisfactorily environmental problems.

Adopting an international political economy perspective, this article has argued that the main problem that international efforts to protect the environment have to address is the issue of the impact of economic globalization.

Economic globalization has in a sense helped to create conditions for the development of policy mechanisms and institutions that will universalize and promote the concept of "sustainable development." Global change is exerting a structuring influence on the redefinition of environmental politics.

However, the kind of sustainable development being promoted seems to represent more the consolidation of a global project of "environmental management" than a real shift away from destructive practices. Globalization is consolidating a market-friendly view of sustainable development, a view that gives priority to the sustainability of "global growth" and to the correction of environmental damage. This tends to be carried out at the expense of the competing alternatives and participative view of sustainable development as stressing not only development but also social equity and decentralized participation.

The "globalization project" has shaped and redefined both the content of environmentalism and environmental policies and structured the international political economy in a way that makes sustainability more difficult to achieve.

DEPENDENCE OF THE AGRICULTURIST ON THE PRODUCE OF FORESTS

These remarks drew attention to the dependence of the agriculturist on the produce of forests and other common lands. This dependence was even more acute in the tribal regions of middle India, where communities of hunter-gatherers, swidden agriculturists, and charcoal iron makers were likewise at the receiving end of the new forest laws. These peoples found an eloquent spokesman in Verrier Elwin (1902-1964), a brilliant Oxford scholar and renegade priest who became the foremost interpreter of adivasi (tribal) culture in India. Elwin was a pioneer of ecological anthropology, whose many works vividly showcased the intimate relationship between the forest world and the life of the adivasi. All tribals, he argued, had a deep knowledge of wild plants and animals; some could even read the great volume of Nature like an "open book." Swidden agriculturists, for whom forest and farm shaded imperceptibly into each other, had an especial bond with the natural world. They liked to think of themselves as children of Dharti Mata, Mother Earth, fed and loved by her.

Elwin's ethnographies are peppered with references to the adivasi's love for the forest. Tragically, the forest and game laws introduced by the British had made them interlopers in their own land. He quotes a member of the tribal group Gond, whose idea of heaven was "miles and miles of forest without any forest guards." As the anthropologist himself wrote in 1941:

The reservation of forests was a very serious blow to the tribesman. He was forbidden to practice his traditional methods of cultivation. He was ordered to remain in one village and not to wander from place to place. When he had cattle he was kept in a state of continual anxiety for fear they should stray over the boundary and render him liable to heavy fines. If he was a Forest Villager he became liable at any moment to be called to work for the Forest Department. If he lived elsewhere he was forced to obtain a license for almost every kind of forest produce. At every turn the Forest Laws cut across his life, limiting, frustrating, destroying his self confidence.

During the year 1933-4 there were 27,000 forest offences registered in the Central Provinces and Berar and probably ten times as many unwhipped of justice. It is obvious that so great a number of offences would not occur unless the forest regulations ran counter to the fundamental needs of the tribesmen. A Forest Officer once said to me: "Our laws are of such a kind that every villager breaks one forest law every day of his life."

Elwin's writings were addressed equally to the colonial state and to the Congress nationalists, who in the 1940s were very much a government-in-waiting. The Congress, however, had not been especially sensitive to the rights of the tribals. But as Elwin reminded them, "the aboriginals are the real swadeshi [indigenous] products of India, in whose presence everything is foreign. They are the ancient people with moral claims and rights thousands of years old. They were here first: they should come first in our regard." He was deeply distressed when a Congress report on tribals followed the British authorities in asking for a ban on shifting cultivation.

Now Elwin's work had shown that, contrary to modernist prejudice, swidden as practiced by the Baiga, the Juang, and other tribes was an ecologically viable system of cultivation. When the nationalists recommended the ban, he wrote angrily that "the forests belong to the aboriginal. I should have thought that anyone who was a Nationalist would at least advocate swaraj [freedom] for the aboriginal!"

The significance of the forest in tribal life is a running theme in Elwin's work. Noting that a majority of tribal rebellions had centered around land and forests, he pleaded for the greater involvement of tribals in forest management in free India. Even if adivasis had no longer any legal rights of ownership, they had considerable moral rights. And as tribals were as much part of the national treasure as forests themselves, there should be an amicable adjustment between forest management and tribal needs. Even where commercial forest operations became necessary, he said, these should be undertaken by tribal cooperatives and not by powerful private contractors. After independence Verrier Elwin became the first foreigner to be granted citizenship of free India. In 1954, he was appointed Adviser on Tribal Affairs to the Government of India (with special reference to the North-east Frontier Agency). He was also to serve on more than one high-level, all-India committee on tribal policy. From his first official appointment until his premature death in 1964, Elwin repeatedly urged a reconsideration of forest policy, such that it might, at last, come to more properly serve tribal needs. In this he had little success for forest management became, if anything, more commercially oriented in independent India.32 Towards the end of his life, the anthropologist wrote with some bitterness of how the victims of government policy were being unfairly blamed for the destruction of forests:

There is constant propaganda that the tribal people are destroying the forest. When this was put to some of the villagers, they countered the complaint by asking how they could destroy the forest. They owned no trucks; they hardly had even a bullock-cart; the utmost that they could carry away was a headload of produce for sale to maintain their families and that too against a license. The utmost that they wanted was wood to keep them warm in the winter months, to reconstruct or repair their huts and carry on their little cottage industries. Their fuel-needs for cooking, they said, were not much, for they had not much to cook. Having explained their own position they invariably turned to the amount of [forest] destruction that was taking place all around them. They asked how the exzamindars [landlords], in violation of their agreements and the forest rules and laws, devastated vast tracts of forest land right in front of officials. They also related how the contractors stray outside the contracted coupes, carry loads in trucks in excess of their authorised capacity and otherwise exploit both the forests and the tribal people.

FOREST CONDITIONS IN DEVELOPING COUNTRIES

Less attention has been paid to forest condition in developing countries because reduction in forest area has attracted more attention. This reduction in density has often been the result of over-exploitation for timber and fuelwood. Overgrazing and repeated bush fires are other significant causes of decrease in density, especially in the dry tropical and non-tropical zones.

Every year, very large areas of savannah woodland and mixed forest-grassland formations are affected by fires set by herders to provide an early flush of green grass when the rains start, particularly in the dry zones of Africa and South America, although no reliable data are available on their extent.

Forests in the humid tropics have also, at times, been affected by large fires, the most serious in recent years having been those associated with the dry weather conditions arising from the El Niño Southern Oscillation (ENSO). In 1997-98, these caused large-scale destruction of logged and secondary forest in Indonesia (particularly Kalimantan, Sumatra and Irian Jaya), Mexico and Central American countries, as well as leading to smoke pollution over wide areas beyond the borders of these countries. The cause of these fires has been burning for land clearance, but it should be noted that the fires are not necessarily in forest; although described as 'forest fires' they are frequently in grassland or on land that has been cleared of forest. It is *estimated* that of the 2 million ha burned in Indonesia in 1997, 150000-200 000 ha may have been in forest. Other serious fires associated with ENSO effects have included that in East Kalimantan, which burned 3.6 million ha in 1983.

Coniferous forests in the humid tropics have often been affected by fires, although it must be noted that fire is frequently necessary to maintain and regenerate them. In the 1980s, the area of pine forest in Honduras and Nicaragua burned annually amounted to some 65 000ha (or about 3.5% of the total pine forest area of these two countries), and widespread fires in natural and artificial tropical pine forests occur in other countries such as Guatemala, Mexico and Indonesia (northern Sumatra).

In the subtemperate and temperate zones of the developing world, fire is also a permanent threat to forests, particularly when they are no longer used by local people for grazing and other purposes. In the 1980s, the average area of forest and other wooded land burned annually was 140 000 ha in the temperate/ subtemperate zones of South America (including southern Brazil). From 1950 to 1990, fires in China are reported to have affected an average of 890 000 ha annually, the most damaging one having been the 'May 6' fire, which burned some 1.85 million ha in the north-eastern province of Heilongjiang in 1987. In the absence of a global statistical fire database, it is difficult to provide an overall estimate of the annual extent of fires in forests and other wooded lands. A very crude estimate for the temperate/subtemperate and humid tropical zones of the developing world (leaving aside the significant dry tropical zone, for which little reliable information exists) would be of the order of 2 million ha of forest and other wooded land annually during the 1980s. Given the lack of sufficient capacity in fire prevention and control in most developing countries, no significant reduction in wooded areas burned is likely to occur in the near future.

Outbreaks of pests and diseases in developing countries are generally reported for those plantations and planted trees where the impact is most apparent. Introduced pests are usually extremely destructive and several of them have had very damaging effects in recent years in the developing world. Examples of introduced insects affecting plantations and planted trees include the *Leu-caena* psyllid, *Heteropsylla cubana,* which has spread into Asia and the Pacific islands and is now extending across Africa; the cypress aphid, *Cinara cupressi,* which is established in eight eastern and southern African countries and is causing heavy mortality among a number of exotic and indigenous species but especially the important plantation species *Cupressus lusitanica,* threatening its future role in the plantation programmes of these countries; and the European woodwasp, *Sirex noctilio,* which has spread into Argentina, Uruguay and southern Brazil, affecting principally *Pinus taeda* but which may become a threat for the large Chilean plantation estate of *Pinus radiata.*

Strategies to control introduced insects from attacking forest trees include regulatory measures, eradication, integrated pest management, and monitoring of population levels and occurrence of potentially harmful species.

Effective monitoring of insect populations will be especially important in view of the large plantation programmes being carried out to produce industrial roundwood, which often rely on one or a few exotic species. Such programmes may be able to afford the investment that is necessary for control measures. Smaller-scale programmes may have to insure against failure due to unexpected attack by insects or diseases by diversifying into several species adapted to the sites and end-uses.

There is also evidence of forest decline due to a combination of biotic and abiotic factors in the developing world, with air pollution likely to play an increasing role in some cases as industrial and transportation infrastructure develops. Examples of such decline include neem *(Azadirachta indica)* in the Sahel, framiré *(Terminalia ivoren-sis)* in Ivory Coast and Ghana, and *Eucalyptus globulus* plantations in Colombia and Peru, while airborne pollution is affecting forests near large cities and industrial areas in China (*e.g. Pinus massoniana* stands near Nanshan). Despite the overall significance of the corresponding damage and losses, surveys of forest decline and diebacks in developing countries remain all too rare.

RECENT ESTIMATES OF GLOBAL FOREST AREA

The following section is based on 1990 baseline figures, updated to 1995, prepared by the Food and Agriculture Organization (FAO) Forest Resource Assessment Programme, reported in and largely drawn from *State of the World's Forests 1997.*

In 1995 forests were estimated to cover 3454 million ha, or 26.6% of the total land area of the world (Greenland and Antarctic excepted). Almost two-thirds of the world's forests were located in seven countries: Russia, Brazil, Canada, USA, China, Indonesia and Zaire; 29 countries had more than half of their land covered by forest, of which 21 were in the tropical belt. However 49 countries, in addition to the many non-forested small island states and territories, had less than 10% of their land covered by forests. Five entire subregions were in this category: North Africa (1.2% of the land area), Near East (1.9%), Temperate Oceania (6.2%), Non-tropical Southern Africa (6.8%) and West Sahelian Africa (7.5%).

The latest information on the distribution of forest and of forest cover change by ecological zones was provided in 1990; the results of the Forest Resources Assessment 2000 were released by FAO in 2000. In 1990, temperate and boreal forests occupied 1.64 billion ha and tropical forests 1.76 billion ha. A breakdown by ecological zone was available only for tropical forests.The figures on tropical forests show that most (88%) are in lowlands; of these, tropical rain forests accounted for 47% of all lowland tropical forests, followed by moist deciduous forest (38%) and dry and very dry formations (15%).

Natural Forests

The forest area estimates described above include undisturbed forests, forests modified by humans through use and management (or 'seminatural' forests) and forests created artificially by humankind (*i.e.* forest plantations) by afforestation or reforestation. (Afforestation is defined as the establishment of a tree crop on an area from which it has always, or for a very long time, been absent. Reforestation is defined as the establishment of a tree crop on forest land.) In most industrialized countries, particularly in continental Europe, forests are being managed in such a way that at management-unit level a continuum exists from low-intensity management, involving natural regeneration, through more intensive methods involving some artificial planting to highly intensive methods with complete planting and cultivation; this makes it difficult to isolate figures for natural forest and plantations. The distinction between natural or seminatural forests and forest plantations can more easily be made for developing countries and some industrialized countries such as New Zealand in which forest plantations have been established using introduced species.

Interest in natural forests, particularly their role in the conservation of biological diversity, has led to efforts to compare forests today with what is thought to be their original character and to give complete protection to areas of forests which have had no, or minimal, human interference. Although there are difficulties in identifying the extent of natural forest, compounded by problems of definition, some information exists that can be used as an indication of broad patterns of natural forests in various regions.

An attempt has been made by the World Wide Fund for Nature (WWF) to quantify the area of forests in western Europe that has been relatively undisturbed by humans or which has retained much of its natural character. Their report distinguishes between 'virgin forest', defined as 'forest ecosystems whose characteristics are determined exclusively by natural location and environmental factors... without human influences present or visible any more', and 'natural and ancient seminatural forests', which 'have not been planted or sown by man for the past two centuries' and 'which continue to have a large number of the natural elements'. The study found that only a small proportion (probably < 1%) of the total forest land in northern and western Europe could be considered as virgin forest, which has arisen since the last glaciation. Almost all was located in Sweden, Finland and Norway, with small areas in Greece, Austria and Switzerland and (according to another author) in France. In eastern Europe, Slovakia and Belarus have considerable areas of virgin forest while Poland and Croatia have small areas. In addition, in northern and western Europe, the WWF report identifies natural and ancient seminatural forests representing 2.1% of the total forest cover (1990) of the 16 countries concerned. There are a further 3 million ha of land in the region in national parks and other protected areas and another 50 000 ha in small forest reserves (mostly for nature conservation and scientific research), whose use is tightly restricted.

The situation in temperate and boreal North America is quite different from that of densely populated Europe and Japan, where use and management of forests for many centuries have left very little of the original forest area untouched. 'Old growth forests', as they are called in North America, still cover extensive areas.

On the lands managed by the US National Park Service alone, old-growth forests covered 1.97 million ha in 1988. Although not a direct measure of the extent of old growth forest, it may be noted that the area of forest included in national parks and other protected areas in North America (USA and Canada) was reported to be nearly 49 million ha in 1990.

The FAO Forest Resource Assessment 1990 did not distinguish between undisturbed and disturbed 'natural' forests in developing countries. However, the Forest Resource Assessment 1980 made estimates of the areas of undisturbed closed forests (primary forests and old secondary forests where there had been no logging for the last 60-80 years) and of closed forests that were included in national parks and other protected areas (thus relatively undisturbed, at least in theory). At that time, these two categories together represented 60% of the total closed forest area in the tropics, a proportion varying from 39% in tropical Asia to 59% in tropical Africa and 69% in tropical America.

These different proportions by region reflected a slower development of large-scale harvesting in tropical America compared with tropical Africa and Asia, and also the fact that, in tropical America, spontaneous colonization did not follow in the wake of logging as systematically as in the two other regions due to lower population pressure. Although the two categories do not match the concept of 'virgin forests' of Europe and of 'old growth forests' of North America, the sum of the two categories nevertheless gives an indication of the amount of forest disturbance (or management, depending on the point of view of the observer) that existed around 1980 in the humid tropics. Although corresponding estimates for 1990 and 1995 are not available, it is likely that the share of undisturbed forests remains higher in the three tropical regions than in Europe and probably in North America too.

DEFORESTATION: GLOBAL IMPACT, CAUSES AND RECOMMENDATIONS

Deforestation is the permanent destruction of forests and woodlands. An issue high on the global environmental agenda for many years, deforestation remains a serious problem today. In the tropics and many other parts of the world, nations continue to lose their natural forests—along with valuable biodiversity, soil and water conservation, and climate regulation these ecosystems provide. Over the past 30 years, the world has lost fully a fifth of all tropical forest cover. While deforestation has stabilized in most developed countries, only a fraction of primary temperate forests still stand. A more pressing issue in these countries today is the condition of the remaining forests. Even though virtually none of their primary forests remain, developed nations continue to allow their commercial exploitation. Meanwhile, pollution and fragmentation endanger ecosystems throughout much of the developed world.

BENEFITS OF FORESTS

Living off the Land

Earth without forests is a picture that most of humankind presently could not conceive. Forests cover much of the planet's land area. They are extremely important to humans and the natural world. For humans they have many aesthetic, recreational, economic, historical, cultural and religious values. Timber and other products of forests are important economically both locally and as exports.

They provide employment for those who harvest the wood or products of the living forest. Herbalists, rubber tappers, hunters and collectors of fungi, nuts, bamboo and berries are able to utilize such resources. Other non-wood forest products come in the form of medicinal compounds, dyes and fabrics. There are many people who are dependent on forestland for their livelihoods. One-third of the world's people depend on wood for fuel as a significant energy source. Surveys in Cameroon, Cote d'Ivoire, Ghana and Liberia found that forest wildlife accounted for 70 to 90% of the total animal protein consumed. Some indigenous peoples are completely dependent on forests. As well as providing a home for some people, the forest environment provides a popular setting for eco-tourism, which includes hiking, camping, bird watching and other outdoor adventure or nature study activities.

Protection from Natural Disasters

Trees protect the soil against erosion, and reduce the risk for landslides and avalanches. They may increase the rate that rainwater recharges groundwater, as well as control the rate that water is released in watersheds. They help to sustain freshwater supplies and therefore are an important factor in the availability of water, one of life's basic needs. When rain falls, some may sink to the ground, some may run off the surface of the land, and flow toward the rivers and some may evaporate. Running water is a major cause of soil erosion. During heavy rains, flooding may occur, filling the waterways with eroded soil. The silt clogs these waterways, cutting off water sources for plants and animals during the dry season. Silt may also fill reservoirs created by dams, reducing its ability and future capacity to generate hydroelectricity and provide irrigation. The removal of forests causes nutrient loss in the soil especially if the period between harvest isn't long enough.

Scientists have recognized that trees can also serve as a tool in the reduction of storm water runoff. The incorporation of trees and other vegetation costs five to ten times less than using solely manmade storm water infrastructures. The leaves on trees keep large quantities of rain and snow from falling to the ground and tree roots absorb excess surface water, thereby stabilizing ground soil. Street trees provide the greatest annual benefit in avoiding storm water runoff by diverting 327 gallons of water compared with the 104 gallons averted by park trees.

Helping the Climate

Researchers have found that trees help the urban ecosystem by decreasing air temperatures. Studies indicate that a 10 percent increase in tree canopy cover results in a one to two degree Fahrenheit reduction in air temperature. In addition a one-degree decrease in temperature will reduce the possibility of smog by 6 percent. Furthermore, increased tree canopy coverage protects urban dwellers from harmful effects of ultraviolet (UV) radiation. Strategic planting of trees can also increase a city's energy efficiency. Research conducted since the mid 1980s has quantified the energy saving potential of urban forests. According to the Energy Information Administration, household heating and cooling cost consumers 180 billion US dollars in 1987. Studies have found that a 25 feet tall tree could save 10 to 25 US dollars annually on these energy costs alone. Because trees release cool vapor into the air during photosynthesis, the need for artificial cooling devices is reduced. In fact, according to one study, the air-conditioning savings from a deciduous tree near a well insulated home ranged from 10 to 15 percent, while an 8 to 10 percent savings was reported during peak cooling periods. Landscape vegetation around individual buildings can also result in heat savings of 5 to 15 percent savings, and cooling savings of 10 to 50 percent.

It has been observed that one of the largest energy fluxes at the earth's surface is that due to evaporation by trees. Heat is absorbed by trees for transpiration of fluid, and later released into the upper atmosphere. The fluid involved is groundwater flowing up the trunk of the tree. Increased annual runoffs from deforested areas in the Amazon support this attribution.

Purification of the Air

Forests affect the climate and are an important source of oxygen (O_2), although they play a lesser role than once thought.. Rain forests serve as an important filter for carbon dioxide (CO_2), a greenhouse gas that contributes to global warming. The Amazon region alone stores at least 75 billion tons of carbon

(C) in its trees. When stripped of its trees, rainforest land soon become useless and inhospitable because the soil lacks the nutrients to support any kind of agriculture. Regeneration of a tropical rainforest may not be possible or, when it can occur, it may take hundreds of years. Research continually reveals that trees benefit urban communities in a number of ways. First with respect to air quality, trees remove damaging pollutants from the atmosphere, and replenish it with O_2. Through the process of transpiration and photosynthesis, trees sequester grams of ozone (O_3), sulfur dioxide (SO_2), nitrogen dioxide (NO_2), and carbon monoxide (CO) every hour, amassing several tons of carbon storage each year. This carbon sequestration process in turn reduces the harmful effects of these noxious gases that cause global warming as well as lung-related ailments. Researchers have also been able to quantify the value of this carbon removal through the use of a carbon storage and sequestration model called UFORE-C. In fact utilizing the figures economists employ to estimate the effect pollutants cost society, one research ecologist was able to compute carbon sequestration into a tangible "dollar-value.".

CAUSES OF DEFORESTATION

There are many causes of deforestation however harmless they may seem. So much damage can be done by even a single chainsaw, because behind those chainsaws are huge companies that care only about demand and profit, and forests are needed to supply this.

Commercial Exploitation

The first and most important cause of deforestation is wood extraction. Wood has always been a primary forest product for human populations and industrial interests. Since wood is an important structural component of any forest, its removal has immediate implications on forest health. Intensive harvests can lead to severe degradation, even beyond a forest's capacity to recover. When the soil has been stripped of its nutrients, farmers move further into the forests in search of new land. Shifting cultivation is one of the most unproductive uses of farmland, and a major cause of degraded land where forests cannot regrow. In eastern India, this agricultural practice called "jhumming," has laid barren previously fertile tracts of the hillside. Timber is one of our most precious as well as used resource. We use it to build our houses, furniture and stock our fireplaces. This heavy demand fuels the destruction of our forests at an unsustainable rate. For every tree that is logged, 27 are killed or damaged in the process. In small doses logging isn't too bad, and over 30 or so years the forest will grow back. But with today's demand for wood the areas that are logged are too large, and this causes permanent destruction of the forests. The impact of the timber trade is generally greater than has been claimed in the past. The North plays a key role in many of the factors leading to forest decline.

Commercial forestry is the leading cause of deforestation in the world's temperate regions. The forces of large global markets for wood and wood products drive the scale of logging activities such as clearcutting. The source of demand is increased consumption by North America and Europe, not population growth. Again, transportation routes have a role, opening up new areas for natural resource exploitation. Privatization of natural resource industries has led to decreased regulations regarding timber harvesting. Multinational corporations dominate trade in wood. Most of these companies were organized in the US. Multinational companies for whom improvement of forest practices is not a priority often export the timber in an unprocessed state out of the country of origin.

Mining for precious resources also plays a major role. There are many forests that hold fair amounts of Earth's resources, such as iron ore, copper, oil and other precious metals. Many mining methods such as strip-mining and strong-force hoses break down the earth and cause major erosion. The mining sites are large and many trees need to be demolished to make way for them. When nothing is left to be mined, there is little chance of the forest growing back because of erosion and the lack of nutrients in the soil, which was churned out during mining.

Development of Degrade Forests

People destroy or degrade forests because, for them, the benefits seem to outweigh the costs. Underlying causes include such issues as poverty, unequal land ownership, women's status, education and to some extent, population. Immediate causes are often concerned with a search for land and resources, including both commercial timber and fuelwood. In many areas, rural households rely solely on fuelwood collected from the forest for their domestic energy supply. The roads that are built into the rainforest encourage and provide access for settling activities. In north-east India road building is often wrecking havoc on the forests. A road is cut through a hill face and the first loss is of the trees along its trace. The debris is thrown down, destroying the trees below, leaving a trail of dead or wilting trees in its wake. This debris enters the valleys, pushing up the levels of streams and rivers, causing siltation and floods.

There are many government agencies with policies that are uncoordinated in nature. Long range planning is not undertaken; and the Amazon is greatly affected by forces outside of the region. Some of the causes of migration to the tropical forests are population growth and political persecution. The settlers clearing and cultivating the land do not have the knowledge and experience of indigenous peoples of the forests and are unable to utilize the land effectively or sustainable. The process of shifting cultivation is accelerated and as a result the forest doesn't have enough time to recover. Tropical rainforests are truly under the assault by humans.

Cattle Ranching

Perhaps the worst culprit of deforestation, at least in the Brazilian Amazon, is cattle ranching, accounting for 38 percent of deforestation in this region. Cattle ranching involves hundreds and thousands of cattle grazing on expansive areas in and near forests. Because the forest soil isn't adapted to these conditions, it isn't long before the area becomes unproductive. So cattle ranchers expand their grazing area, leading to more destruction. After the grazed land is left, the forest is very unlikely to grow back due to the stripping of the soil. Cattle ranching each year in this manner destroys an estimated 5700 square miles of rainforest alone. The Brazilian government subsidizes some of the cattle ranches that exist on converted forestland. The land is unproductive. Much of the demand for the beef comes from the fast food hamburger market, which is more concerned with quantity, than quality farm raised meat.

Agriculture, Ecosystem and Forests

Monocultural forestry simplifies the ecosystem, leaving it vulnerable to disease and other environmental factors. In the tropical forests of the world, the clearing of land for agriculture and livestock are the primary activities resulting in deforestation. The main cause is unequal distribution of land. 4.5 percent of Brazil's landowners own 81 percent of the country's farmland, and 70 percent of the rural households are landless. It seems that these conditions cause people to encroach on, penetrate and modify the forests. Governments have an important role in these processes. In many countries in Asia and Africa, where family farms are still prevalent, the breakdown of large joint-families is causing uneconomic divisions to existing farms. These inefficiencies, in turn, put pressure on farmers to sell their land for development, and it turns whole farming comminutes into new developments.

Poverty and Inequality

Another cause of the ecological crisis of the present is social inequality. Gender inequality is one of the more powerful forces at work, which exists in virtually all of human cultures. The natural world is often portrayed as feminine, in terms such as "Mother Nature," "virgin forest," "exploitation" and "rape of the valley" that are used to describe elements and uses of nature and serve to perpetuate this harmful attitude. Human society's attitude with regard to the status of women makes an important contribution to environmental degradation and deforestation.

Although it is easy to assume a strong connection between population growth and deforestation, some research indicates that the problem is more complex. It involves non-demographic mechanisms resulting from credit and capital market failures, lack of suitable mediating institutions securing property rights, wretched poverty, uneven land distribution, consumption patterns in developed countries, greedy multinational companies, ignorance and bad management by colonist or frontier land, and so forth.

One of the underlying causes of human exploitation and consumption of forests and other natural resources is human tradition and beliefs. One source of such belief is Christianity, whose dominance in the Americas and Europe has important consequences for natural resources. Christian's attitudes are of anthropocentrism. The dominant power on the planet is humankind. The first human, Adam, gave the animals their names, shows this kind of dominance. Modern science and technological changes began in the name of Christianity. These beliefs created the attitudes, traditions and activities that enable us to be responsible for the destruction of nature that is occurring in the forests and the rest of the world. It is then possible that if certain Christian beliefs were different, human attitude towards nature would be that of conservation not exploitation. Such contrasting beliefs could include a god or gods that exist on Earth or even in trees, or that humans are reincarnated as plants and animals.

Bibliography

A K Bhattacharya: Public People Private Partnership for Sustainable Forest Development, Concept, Publication, Delhi, 2007.

Ajay S. Rawat: Forest Management in Kumaon Himalaya, Indus, 2000.

Ajoy Kumar Bhattacharya: Community Participation and Sustainable Forest Development: Global Perspective, Concept, Publication, Delhi, 2001

D.D. Tewari: Management of Nontimber Forest Product Resources of India : An Analysis of Forest Development, International Book, Delhi, 2008.

Hemant Kumar Gupta: Joint Forest Management: Policy, Participation and Practice in India, International Book, Delhi, 2006.

J B Lal: Economic Tools and Institutional Framework for Forest Development Planning, International Book, 2005.

K.D. Singh: Community Based Forest Management With Micro Enterprise for Poverty Reduction, , Bishen Singh Mahendra Pal Singh, Publication, Dehradun, 2009.

K.M. Bhat, K.K.N. Nair, K.V. Bhat, E.M. Muralidharan and J.K. Sharma: Quality Timber Products of Teak from Sustainable Forest Management, Kerala Forest Research Institute, 2005.

Klaus Seeland and Franz Schmithusen: Indigenous Knowledge, Forest Management and Forest Policy in South Asia, D K Printworld, 2003.

L. Robin Singh: Social Forestry and Forest Development Planning, Daya, Publication, Delhi, 1997.

M R K Jerram: Text Book on Forest Management, International Book, Delhi, 2006.

M.M. Deka: Joint Forest Management in Assam, Daya, Publication, Delhi, 2002.

M.P. Shiva and S.K. Verma: Approaches to Sustainable Forest Management and Biodiversity Conservation, International, 2002.

M.R.K. Jerram: A Text-Book on Forest Management: In Agriculture, Horticulture and Forestry, Asiatic Publication, Delhi, 2006.

Madhusudan Bandi: Tribals and Community Forest Management, Rawat Publications, Delhi, 2013.

Mitali Chatterjee and S.B. Roy: Reflections on Gender Issues in Joint Forest Management, Inter-India, 2000.

N.H. Ravindranath and P. Sudha: Joint Forest Management in India: Spread, Performance and Impact, Universities Press, 2004.

Nandini Sundar: Branching Out: Joint Forest Management in India, Roger Jeffery and Neil Thin, OUP, 2001.

P. Bhattacharya, A.K. Kandya and K.N. Krishna Kumar: Joint Forest Management in India, Aavishkar, Publication, Delhi, 2008.

P.C. Kotwal and M.D. Om Prakash: Criteria and Indicators for Sustainable Forest Management, International Book Distributors, 2007.

Raj Kumar Sen: Forest Management and Sustainable Development, Deep and Deep, Publication, Delhi, 2012.

Ram Prakash: Forest Management, International Book, Delhi, 2006.

S Dasgupta and D Debnath: Community Forest Management in Tribal States of India, International Book, Delhi, 2008.

S. Balaji: Joint Forest Management for Bio-Diversity Enhancement, Bishen Singh Mahendra Pal Singh, Publication, Dehradun, 2004.

S. John Joseph and B.S. Nagarajan: A Paradigm Shift in Forest Management : (Based on a Study into the Changing Profile of Sirumalai Forest Hills, Dindigul District, Tamil Nadu), International Book, Delhi, 2004.

S.K. Ghosh and Rita Singh: Social Forestry and Forest Management, Global Vision, Publication, 2003.

S.S. Negi: Forest Management in India, Bishen Singh Mahendra Pal Singh, Publication, Dehradun, 2011.

S.S. Sagwal: Dictionary of Forest Management: Includes Economics Fire Range and Wildlife, APC, 1996.

Steven E. Franklin: Remote Sensing for Sustainable Forest Management, Routledge, Publication, 2010.

Index